Calculus
of Several Variables
and
Differentiable Manifolds

Calculus of Several Variables and Differentiable Manifolds

Carl B. Allendoerfer
UNIVERSITY OF WASHINGTON

Macmillan Publishing Co., Inc.
New York
Collier Macmillan Limited
London

Printed in the United States of America

Macmillan Publishing Co., Inc.
866 Third Avenue, New York, New York 10022

Collier-Macmillan Canada, Ltd., Toronto, Ontario

Library of Congress Cataloging in Publication Data

Allendoerfer, Carl Barnett, 1911–
Calculus of several variables and differentiable manifolds.

1. Calculus. 2. Functions of several complex variables. 3. Differentiable manifolds. I. Title.
QA303.A44 515'.94 73–14790
ISBN 0-02-301840-2

Printing: 1 2 3 4 5 6 7 8 Year: 4 5 6 7 8 9 0

Preface

This volume is a revised version of lecture notes prepared for a first-year graduate course at the University of Washington that I have given from time to time. The purpose of the course is to introduce the students to differentiable manifolds and the various structures on them so that they may use this material in later courses in Differential Geometry, Lie Groups, Functions of Several Complex Variables, and the like.

Experience with the course has shown the necessity of preliminary material on Advanced Calculus, Differential Equations and Multilinear Algebra which apparently are not covered in a satisfactory fashion in many undergraduate departments. Moreover, students cannot appreciate the theory of differentiable manifolds in a proper fashion unless they have an elementary knowledge of the geometry of surfaces imbedded in Euclidean space. For the definitions of the structures on a differentiable manifold are all abstractions from those occuring naturally in surface theory. One reason why this background is so often neglected is that the needed material is scattered through fat books on Advanced Calculus, Differential Equations, Differential Geometry and Linear Algebra; and so it is difficult to dig it out in a reasonable time. For this reason, Chapters 1 to 5 have been prepared as a summary of the needed background and as an introduction to the manifold theory of Chapters 6 and 7.

Since the subject matter of this volume has been worked over intensely by many mathematicians during the past 35 years, there is little room for innovation here. I have drawn extensively on the works of others, but as the course has developed through the years, I have lost track of the precise sources from which I have drawn specific materials. The major ones are listed in the references at the ends of Chapters 2 and 7. In addition, I give grateful acknowledgement for what I learned from my teachers L. P. Eisenhart, A. E. H. Love, W. Mayer and T. Y. Thomas, and from my former colleagues A. Nijenhuis and S. Kobayashi. I have, however, rethought the material and have attempted to present it in a coherent notation and in a unified fashion.

A possibly novel feature is the interplay between Differential Equations and Advanced Calculus. By the use of an old, but neglected, method the differential equation form of the Frobenius Theorem is proved (even though Frobenius would scarcely recognize his theorem). Then this is used to prove the inverse and implicit function theorems without the usual anguish, and later to obtain the other forms of the Frobenius Theorem. Thus, the dirty work is done only once, and the other results follow easily.

Prerequisites for reading this volume are an understanding of the topological portions of Advanced Calculus including the uniform convergence of series, standard Linear Algebra and mathematical maturity at the level of an undergraduate major. Knowledge of elementary Algebraic Topology would be useful but is not essential.

I am very grateful to all the unnamed mathematicians whose ideas I have utilized, and to my students whose questions have helped me immensely in improving my exposition and in correcting my errors.

Carl B. Allendoerfer
August 1973

Table of Contents

		Page
Chapter 1	DIFFERENTIABLE FUNCTIONS	
1.1	Euclidean Space, R^n	1
1.2	Functions from R^n to R^m	1
1.3	Examples	2
1.4	Continuity	2
1.5	Differentiability	3
1.6	Partial Derivatives	4
1.7	Interpretation	7
1.8	Directional Derivatives	9
1.9	Chain Rule	9
1.10	Mean Value Theorem for $f: R^n \to R^1$	12
1.11	Mean Value Theorem for $f: R^n \to R^m$	14
1.12	Taylor's Theorem	15
1.13	Integration	18
Chapter 2	EXISTENCE THEOREMS FOR DIFFERENTIAL EQUATIONS	
2.1	Introduction	23
2.2	The Basic Theorem	23
2.3	The Picard Proof by Successive Approximations	24
2.4	Contraction Map Proof	29
2.5	Refinements	32
2.6	Approximate Solutions	32
2.7	Linear Equations	34

		Page
Chapter 2	EXISTENCE THEOREMS FOR DIFFERENTIAL EQUATIONS (Continued)	
2.8	Dependence of Solutions on Initial Values and Parameters	34
2.9	Partial Differential Equations - First Problem	39
2.10	The Problem of Section 2.9 When D Is Not Simply Connected	44
2.11	First Form of the Frobenius Theorem	47
Chapter 3	THEOREMS ABOUT DIFFERENTIABLE FUNCTIONS	
3.1	Introduction	54
3.2	First Proof of the Inverse Function Theorem	54
3.3	Second Proof of the Inverse Function Theorem	57
3.4	Implicit Function Theorem	58
3.5	Second Proof of the Implicit Function Theorem	63
3.6	Local Submanifolds of R^n	64
3.7	Tangent and Normal Vectors for a Submanifold of R^n	68
3.8	Extreme Values of $f: R^n \to R^1$	70
3.9	Constrained Maxima and Minima	71
3.10	Diagonalization of a Symmetric Matrix	75
3.11	Frobenius Theorem (Second Form)	78
3.12	Third Form of the Frobenius Theorem (Preliminary Version)	82
Chapter 4	THE GEOMETRY OF SUBMANIFOLDS	
4.1	Introduction	89
4.2	Definition of a Curve in R^n	89

		Page
Chapter 4	THE GEOMETRY OF SUBMANIFOLDS (Continued)	
4.3	Arc Length	91
4.4	Frenet Equations of a Curve	92
4.5	Solution of the Frenet Equations	95
4.6	Hypersurfaces in R^n	97
4.7	Arc Length on a Hypersurface	98
4.8	The Frenet Equations of a Hypersurface	100
4.9	The Coefficients $\Gamma^{\gamma}_{\alpha\beta}$	102
4.10	Covariant Derivatives of Vectors	105
4.11	The Principal Curvatures of a Hypersurface	109
4.12	The Gauss and Codazzi Equations	113
4.13	Volume (area) of a Hypersurface	115
4.14	The Spherical Image of M^{n-1} in R^n	118
Chapter 5	MULTILINEAR ALGEBRA	
5.1	Introduction	123
5.2	Dual Vectors	123
5.3	Tensors on V	124
5.4	The Double Dual Vector Space V^{**}	127
5.5	Products of Tensors	128
5.6	The Contraction Operator	130
5.7	The Exterior Algebra of Covectors	131
5.8	Exterior Products of Covectors	134
5.9	Orientation of V	135

		Page
Chapter 6	DIFFERENTIABLE MANIFOLDS	
6.1	Introduction	142
6.2	Functions on M	143
6.3	Tangent Vectors on M	144
6.4	Tangent Vectors to a C^p Manifold	147
6.5	The Tangent Bundle $T(M)$	149
6.6	Vector Fields	151
6.7	Lie Bracket	152
6.8	Cotangent Vectors on M	152
6.9	Tensor Fields and p-forms on M	155
6.10	Exterior Derivatives of p-forms on M	156
6.11	Frobenius Theorem (Third Form)	159
6.12	Mappings	163
6.13	Sard's Theorem	165
6.14	Connections and Covariant Derivatives	168
6.15	The Covariant Exterior Derivative	171
6.16	Parallel Displacement	171
6.17	Orientation of M	173
Chapter 7	INTEGRATION OF FORMS ON MANIFOLDS	
7.1	Integration of Forms in R^p	180
7.2	Stokes' Theorem for I^p	181
7.3	Stokes' Theorem for Cubical Singular Chains	183
7.4	Consequences of Stokes' Theorem	185
7.5	Poincare Lemma	188
7.6	Partition of Unity	191

			Page
Chapter 7	INTEGRATION OF FORMS ON MANIFOLDS (Continued)		
7.7	Change of Variables in a Multiple Integral		193
7.8	Integration on Manifolds		194
7.9	Manifolds with Boundary		197
7.10	Stokes' Theorem on Manifolds with Boundary		200
7.11	Proof of Change of Variable Theorem for Multiple Integrals		202
7.12	Integration of Forms on Submanifolds		205
7.13	Riemannian Metric		205
7.14	Integration on Riemannian Manifolds		207
7.15	Classical Forms of Stokes' Theorem		212
	Index		221

Chapter 1

Differentiable Functions

1.1 Euclidean Space, R^n. In this section we review essential concepts and establish notation to be used hereafter.

R^n is a vector space over the real numbers of dimension n. Its "usual basis" is:

$$e_1 = (1, 0, \ldots, 0)$$

$$e_2 = (0, 1, \ldots, 0)$$

$$---$$

$$e_n = (0, 0, \ldots, 1)$$

In terms of this basis, a point x of R^n has the components (coordinates) $x^1, \ldots, x^n$ which are real numbers.

An inner product $\langle x,y \rangle$ is defined in terms of the usual basis by

$$\langle x,y \rangle = \sum_i x^i y^i$$

The norm of a vector x is given by

$$|x| = \sqrt{\langle x,y \rangle}$$

This norm satisfies the triangle inequality:

$$|x + y| \leq |x| + |y|$$

The standard topology is defined by taking as a base for the open sets the open balls $\{x: |x - a| < r\}$. R^n is complete in the sense that every Cauchy sequence converges.

1.2 Functions from R^n to R^m. The usual notation for such a function is $f\colon R^n \to R^m$, but other notations will be used.

If we choose a basis in R^n, a point has coordinates $(x^1, \ldots, x^n)$ and $f\colon R^n \to R^1$ can be written $f(x^1, \ldots, x^n)$ or simply $f(x)$. Special cases of such functions are the <u>projection functions</u>.

$$\pi^i(x^1, \ldots, x^n) = x^i \quad (i = 1, \ldots, n)$$

If we choose a basis in R^m, we can define the <u>component functions</u> of $f\colon R^n \to R^m$ by

$$f^\alpha(x) = \pi^\alpha[f(x)] = (\pi^\alpha \circ f)\,(x); \; \alpha = 1 \ldots m.$$

Each $f^\alpha(x)$ is a function $R^n \to R^1$. Given $f\colon R^n \to R^m$, m component functions are defined, and given m functions $f^\alpha(x)\colon R^n \to R^1$ and a basis for R^m, a function $f\colon R^n \to R^m$ is defined. The component functions depend on the choice of a basis in R^m.

1.3 <u>Examples</u> (where the usual bases are assumed)

(a) $f\colon R^1 \to R^1 \qquad y = f(x)$

(b) $f\colon R^2 \to R^1 \qquad z = f(x,y)$

(c) $f\colon R^1 \to R^2 \qquad \begin{cases} x^1 = f^1(t) \\ x^2 = f^2(t) \end{cases}$

(d) $f\colon R^2 \to R^2 \qquad \begin{cases} x^1 = f^1(u^1,u^2) \\ x^2 = f^2(u^1,u^2) \end{cases}$

(e) $f\colon R^2 \to R^3 \qquad \begin{cases} x^1 = f^1(u^1,u^2) \\ x^2 = f^2(u^1,u^2) \\ x^3 = f^3(u^1,u^2) \end{cases}$

1.4 <u>Continuity</u>.

<u>Definition</u>. $f\colon R^n \to R^m$ is continuous at $\underline{a} \in R^n$ iff for each open set $U[f(a)]$ in R^m there is an open set $V(a)$ in R^n

such that

$$f[V(a)] \subset U[f(a)]$$

<u>Theorem 1</u>. The projection functions $\pi^\alpha(x)$ are continuous at every point of R^n.

<u>Proof</u>. $|x - a| = [(x^1 - a^1)^2 + \ldots + (x^n - a^n)^2]^{1/2}$

So $|x^i - a^i| < \varepsilon$ if $|x - a| < \varepsilon$.

<u>Theorem 2</u>. If $f\colon R^n \to R^m$ is continuous at $\underline{a} \in R^n$, then the component functions $f^\alpha(x) = (\pi^\alpha \circ f)(x)$ are continuous at $\underline{a}$.

<u>Proof</u>. Since π^α and f are continuous, the composite $\pi^\alpha \circ f$ is continuous.

1.5 <u>Differentiability</u>. In the usual calculus course a function $f(x)$ is called differentiable at $x = a$ if

$$\lim_{h\to 0} \frac{f(a + h) - f(a)}{h} = f'(a)$$

exists. This is clearly equivalent to:

$$\lim_{h\to 0} \frac{f(a + h) - f(a) - h\, f'(a)}{h} = 0$$

As a generalization we state the definition:

<u>Definition</u>. $f\colon R^n \to R^m$ is differentiable at $\underline{a} \in R^n$ iff there is a linear transformation $L_a\colon R^n \to R^m$ such that

$$\lim_{h\to 0} \frac{|f(a + h) - f(a) - L_a(h)|}{|h|} = 0$$

where h is a vector of R^n.

If we choose bases in R^n and R^m, and let L_i^α be the matrix of L_a relative to these bases, this definition is equivalent to (see exercise 2):

$$\lim_{h\to 0} \frac{f^\alpha(a + h) - f^\alpha(a) - \sum_i L_i^\alpha h^i}{|h|} = 0$$

Hence we have the result:

Theorem 3. $f: R^n \to R^m$ is differentiable at $\underline{a} \varepsilon R^n$ iff its component functions $f^\alpha(x)$ are differentiable at $\underline{a}$.

Definition. The linear transformation L_a is called the differential of f at $\underline{a}$. It is frequently denoted by df_a.

Definition. The norm $|df|$ of df is the greatest lower bound of all numbers C such that $|df(h)| \leq C|h|$. Compare exercise 13.

Theorem 4. If $f: R^n \to R^m$ is differentiable at $\underline{a}$, then the linear transformation L_a is unique.

Proof. Suppose that there is another linear transformation K_a such that

$$\lim_{h\to 0} \frac{|f(a+h) - f(a) - K(h)|}{|h|} = 0$$

Then

$$\lim_{h\to 0} \frac{|L_a(h) - K_a(h)|}{|h|} \leq$$

$$\lim_{h\to 0} \frac{|L_a(h) - f(a+h) + f(a)|}{|h|} + \lim_{h\to 0} \frac{|f(a+h) - f(a) - K_a(h)|}{|h|} = 0$$

Now put $h = tx$ where x is an arbitrary fixed vector in R^n, and $x \neq 0$. Then

$$0 = \lim_{t\to 0} \frac{|L_a(tx) - K_a(tx)|}{|tx|} = \frac{|L_a(x) - K_a(x)|}{|x|} \quad \text{for all } x.$$

Hence $L_a = K_a$.

1.6 Partial Derivatives. Choose bases in both R^n and R^m. Then the partial derivatives $\partial f^\alpha/\partial x^i$ $(\alpha = 1,\ldots,m,$ $i = 1,\ldots,n)$ are defined by:

$$\frac{\partial f^\alpha}{\partial x^i}(a) = \lim_{t\to 0} \frac{f^\alpha(a + te_i) - f^\alpha(a)}{t}$$

<u>Theorem 5</u>. If $f: R^n \to R^m$ is differentiable at $\underline{a}$, the partial derivatives of f at $\underline{a}$ exist and their matrix is the matrix of L_a with respect to the chosen bases.

<u>Proof</u>. By hypothesis there exists L_j^α such that

$$\lim_{h \to 0} \frac{f^\alpha(a + h) - f^\alpha(a) - \sum_j L_j^\alpha h^j}{|h|} = 0$$

Put $h = te_i$. Then $h^j = \begin{cases} t & j = i \\ 0 & j \neq i \end{cases}$

So it follows that

$$\lim_{t \to 0} \frac{f^\alpha(a + te_i) - f^\alpha(a) - L_i^\alpha t}{t} = 0$$

or

$$\lim_{t \to 0} \frac{f^\alpha(a + te_i) - f^\alpha(a)}{t} = L_i^\alpha$$

Therefore $\partial f^\alpha / \partial x^i = L_i^\alpha$

<u>Definition</u>. The matrix of partial derivatives is called the Jacobian matrix and is written

$$\frac{\partial(f^1, \ldots, f^m)}{\partial(x^1, \ldots, x^n)} = J[f(x)]$$

When $n = m$, the determinant of $J[f(x)]$ is called the Jacobian of f.

<u>Definition</u>. The function $f: R^n \to R^m$ is continuously differentiable at $\underline{a}$ iff it is differentiable in $U(a)$ and if each partial derivative is continuous at $\underline{a}$. Such a function is said to be of class C^1. If all r^{th} order partial derivatives exist and are continuous, the function is of class C^r. If all partial derivatives exist and are continuous, the function is of class

C^∞ or "smooth". A C^∞ function is analytic, C^ω, if the Taylor series expansion of each component function converges to the function.

Example 1. The function $f(x) = \begin{cases} e^{-1/x^2} & x \neq 0 \\ 0 & x = 0 \end{cases}$

is C^∞ but not C^ω.

Example 2. The function $f(x) = \begin{cases} x^2 \sin 1/x & x \neq 0 \\ 0 & x = 0 \end{cases}$

is differentiable at $x = 0$, but is not of class C^1 at 0.

Remark. The converse of Theorem 5 is false.

Example 3. $f(x,y) = \begin{cases} \dfrac{2xy^2}{x^2 + y^4} & (x,y) \neq (0,0) \\ 0 & (x,y) = (0,0) \end{cases}$

has $\frac{\partial f}{\partial x}$ and $\frac{\partial f}{\partial y}$ defined at $(0,0)$ but is not differentiable there.

The following partial converse of Theorem 5, however, is true.

Theorem 6. If all partial derivatives $\frac{\partial f^\alpha}{\partial x^i}$ exist in an open ball $U(a)$ and are continuous at $\underline{a}$, then f is differentiable at $\underline{a}$.

Proof. We seek to show that

$$\lim_{h \to 0} \frac{f^\alpha(a + h) - f^\alpha(a) - \sum_i (\partial f^\alpha/\partial x^i)_a h^i}{|h|} = 0$$

Define

$$g_i^\alpha(a + h) = f^\alpha(a^1 + h^1, \ldots, a^i + h^i, a^{i+1}, \ldots, a^n)$$
$$- f^\alpha(a^1 + h^1, \ldots, a^{i-1} + h^{i-1}, a^i, \ldots, a^n)$$

Then $f^\alpha(a + h) - f^\alpha(a) = \sum_{i=1}^{n} g_i^\alpha(a + h)$

From the mean value theorem of one-variable calculus:

$$g_i^\alpha(a+h) = \frac{\partial f^\alpha}{\partial x^i}(a^1 + h^1, \ldots, a^{i-1} + h^{i-1}, a^i + s_\alpha h^i, a^{i+1}, \ldots, a^n)h^i$$

where $0 < s_\alpha < 1$.
Then

$$\lim_{h\to 0} \frac{f^\alpha(a + h) - f^\alpha(a) - \sum_i (\partial f^\alpha/\partial x^i)_a h^i}{|h|} =$$

$$\lim_{h\to 0} \frac{\sum_i \left[\frac{\partial f^\alpha}{\partial x^i}(a^1 + h^1, \ldots, a^{i-1} + h^{i-1}, a^i + s_\alpha h^i, a^{i+1}, \ldots, a^n) - \left(\frac{\partial f^\alpha}{\partial x^i}\right)_a\right] h^i}{|h|}$$

which is zero, since $\frac{\partial f^\alpha}{\partial x^i}$ are continuous at $x = a$ and $\frac{h^i}{|h|}$ are bounded.

Note, however, that f may be differentiable even though its partial derivatives are not continuous. See exercise 9.

1.7 Interpretation. In elementary calculus we know that the equation of the tangent line to $y = f(x)$ at $(a, f(a))$ is

$$y = f(a) + f'(a)\,(x-a)$$

Also

$$f(x) - [f(a) + f'(a)\,(x-a)] = [f(x) - f(a)] - f'(a)(x-a)$$

$$= \left[\frac{f(x) - f(a)}{x - a} - f'(a)\right](x-a)$$

Hence $$\lim_{x\to a} \frac{|f(x) - [f(a) + f'(a)(x-a)]|}{x - a} = 0$$

We interpret this by saying that $f(a) + f'(a)(x - a)$ is a "good" approximation to $f(x)$ near $\underline{a}$, or that $f(x)$ is approximated near $x = a$ by the affine function $f(a) + f'(a)(x-a)$ whose graph is a straight line.

Similarly for a differentiable function $f: R^n \to R^m$, we can define the affine function:

$$A(x-a) = f(a) + df_a\,(x - a)$$

This function is similarly a "good" approximation to f near $\underline{a}$ in the sense that

$$\lim_{x \to a} \frac{|f(x) - A(x - a)|}{|x - a|} = 0$$

For if we write $h = x - a$ and substitute we get

$$\lim_{h \to 0} \frac{|f(a + h) - f(a) - df_a(h)|}{|h|} = 0$$

Example 4. For $f: R^2 \to R^1$ we may write $z = f(x,y)$. Then the affine function

$$f(a,b) + \left(\frac{\partial f}{\partial x}\right)_{a,b} (x - a) + \left(\frac{\partial f}{\partial y}\right)_{a,b} (y - b)$$

is a "good" approximation to $f(x,y)$ near (a,b). The equation of the tangent plane at (a,b) is

$$z = f(a,b) + \left(\frac{\partial f}{\partial x}\right)_{a,b} (x - a) + \left(\frac{\partial f}{\partial y}\right)_{a,b} (y - b)$$

This shows that near a point $\underline{a} \in R^n$, a differentiable function behaves very much like an affine function. Hence we can conjecture local properties of differentiable functions from the known (elementary) properties of affine functions.

1.8 <u>Directional Derivatives</u>. Here we consider $f\colon R^n \to R^1$.

<u>Definition</u>. The derivative of f at $\underline{a} \in R^n$ in the direction $v \in R^n$ is (assume $v \neq 0$):

$$\left(\frac{df}{dv}\right)_a = \lim_{t\to 0} \frac{f(a + tv) - f(a)}{t}$$

<u>Theorem 7</u>. If f is differentiable at $\underline{a}$, then $(df/dv)_a$ exists for every v and

$$\left(\frac{df}{dv}\right)_a = df_a(v) = \sum_i \left(\frac{\partial f}{\partial x^i}\right)_a v^i$$

<u>Proof</u>.

$$\frac{|f(a + tv) - f(a) - df_a(tv)|}{|tv|} = \frac{1}{|v|}\left|\frac{f(a + tv) - f(a)}{t} - df_a(v)\right|$$

The limit as $t \to 0$ of the left side is zero. Hence from the right side $\left(\frac{df}{dv}\right)_a = \lim_{t\to 0} \frac{f(a + tv) - f(a)}{t} = df_a(v)$

<u>Remark</u>. The formula $\left(\frac{df}{dv}\right)_a = \sum_i \left(\frac{\partial f}{\partial x^i}\right)_a v^i$ does not necessarily give the directional derivative of f unless f is differentiable at $\underline{a}$. The formula may make sense even when the directional derivative does not exist. See exercise 11.

1.9 <u>Chain Rule</u>. We observe that for the affine function A which approximates $f\colon R^n \to R^m$ at $\underline{a}$:

$$A\,(x - a) = f(a) + df_a\,(x - a)$$

we have: $dA_a = df_a$

for
$$\lim_{h\to 0} \frac{|A(x + h - a) - A(x - a) - df_a(h)|}{|h|}$$

$$= \lim_{h\to 0} \frac{|f(a) + df_a(x + h - a) - f(a) - df_a(x - a) - df_a(h)|}{|h|} = 0$$

If we have a differentiable function $g: R^m \to R^p$, it is approximated at $f(a)$ by the affine function

$$B[y - f(a)] = (g \circ f)(a) + dg_{f(a)}[y - f(a)]$$

where y is a point of R^m.

The composition $(B \circ A)(x - a)$ is

$$(B \circ A)(x - a) = (g \circ f)(a) + dg_{f(a)}[df_a(x - a)]$$

$$= (g \circ f)(a) + (dg_{f(a)} \circ df_a)(x - a)$$

and its differential is $dg_{f(a)} \circ df_a$. Since these affine functions are "good" approximations to the original functions, it is a reasonable conjecture that the differential of $g \circ f$ is $dg_{f(a)} \circ df_a$. This is indeed true and is proved in Theorem 8.

<u>Theorem 8</u>. <u>Chain Rule</u>

Let $f: R^n \to R^m$ be differentiable at $\underline{a} \in R^n$

$g: R^m \to R^p$ be differentiable at $f(a) \in R^m$

Write $g \circ f : R^n \to R^p$

Then $g \circ f$ is differentiable at $\underline{a} \in R^n$ and

$$d(g \circ f)_a = dg_{f(a)} \circ df_a$$

<u>Proof</u>. Let the vector ε be defined by the equation

$$f(a + h) - f(a) - df_a(h) = \varepsilon|h|$$

Then by hypothesis $\lim_{h\to 0} |\varepsilon| = 0$

Similarly let the vector η be defined by the equation

$$g[f(a + h)] - g[f(a)] - dg_{f(a)}[f(a + h) - f(a)] = \eta|f(a + h)-f(a)|$$

Then by hypothesis $\lim_{h\to 0} |\eta| = 0$

So

$$g[f(a+h)] - g[f(a)] - dg_{f(a)}[df_a(h) + \varepsilon|h|] = \eta|f(a+h) - f(a)|$$

or

$$g[f(a+h)] - g[f(a)] - (dg_{f(a)} \circ df_a)(h) =$$

$$dg_{f(a)}(\varepsilon|h|) + \eta|f(a+h) - f(a)|$$

and

$$\lim_{h\to 0} \frac{|(g \circ f)(a+h) - (g \circ f)(a) - (dg_{f(a)} \circ df_a)(h)|}{|h|} \le$$

$$\lim_{h\to 0} |dg_{f(a)}(\varepsilon)| + \lim_{h\to 0} |\eta| \frac{|f(a+h)| - f(a)|}{|h|}$$

$$= 0 + \lim_{h\to 0} |\eta| \frac{|df_a(h) + \varepsilon|h||}{|h|}$$

$$= \lim_{h\to 0} |\eta| \frac{|df_a(h)|}{|h|} + 0$$

$= 0$ since $\dfrac{|df_a(h)|}{|h|}$ is bounded. (See exercise 13)

Corollary. If f and g are differentiable as in the theorem,

$$J_a(g \circ f) = J_{f(a)}g \times J_a f \quad \text{(matrix multiplication)}$$

Remark. If f and g are not differentiable, their partial derivatives may still exist. The right side of the equation in the corollary is then meaningful, but the equality may not be true. It is true, however, if the partial derivatives are continuous.

1.10 Mean Value Theorem for $f: R^n \to R^1$.

There are several theorems that are generalizations of the mean value theorem of elementary calculus. Here we give the simplest of these for functions $f: R^n \to R^1$.

Lemma. Let $g(t) = f(a + th)$ and suppose that f is differentiable at $a + th$. Then

$$\frac{dg}{dt} = df_{(a + th)}(h)$$

Proof. Apply the chain rule to $f(a + th)$

Theorem 9. Mean Value Theorem. Let $f: R^n \to R^1$ be differentiable at every point of the line segment $a + th$ for $0 \leq t \leq 1$. Then there exists a real number $s \in (0,1)$ such that

$$f(a + h) - f(a) = df_{(a + sh)}(h)$$

Proof. Consider $g(t) = f(a + th)$ as in the lemma. Then by the mean value theorem of elementary calculus, there exists an s such that

$$g(1) - g(0) = (dg/dt)_s$$

Hence

$$f(a + h) - f(a) = df_{(a + sh)}(h)$$

This may also be written:

$$f(x_2) - f(x_1) = df_\xi(x_2 - x_1)$$

where ξ is a point on the segment joining x_1 and x_2.

A related theorem is frequently useful in applications:

Theorem 10. Let $f: R^n \to R^1$ be of class C^p $(p \geq 1)$ in an open convex set U, and let x_1 and x_2 be two points of U. Then in $U \times U$ there exist functions $g_i(x_1, x_2)$ such that

(1) $f(x_2) - f(x_1) = \sum_i (x_2^i - x_1^i) g_i\ (x_1, x_2)$

(2) $g_i(x_1, x_2)$ are of class C^{p-1} in $U \times U$ and

(3) $g_i(x,x) = (\partial f/\partial x^i)_x$. If f is C^∞ or C^ω so are g_i.

Proof. Write $g_i(x_2, x_1) = \int_0^1 \frac{\partial f}{\partial x^i} [(x_2 - x_1)t + x_1]dt$

Then $g_i(x_2, x_1)$ are of class C^{p-1} and $g_i(x,x) = \left(\frac{\partial f}{\partial x^i}\right)_x$

Hence

$$\sum_i (x_2^i - x_1^i) g_i(x_1, x_2) = \int_0^1 \sum_i (x_2^i - x_1^i) \frac{\partial f}{\partial x^i} [(x_2 - x_1)t + x_1]dt$$

$$= \int_0^1 \frac{df}{dt} [(x_2 - x_1)t + x_1]\ dt$$

$$= f(x_2) - f(x_1)$$

Remark. If we put $x_2 = x$ and $x_1 = a$ where $\underline{a}$ is fixed, this theorem can be stated: Let $f: R^n \to R^1$ be of class C^p $(p \geq 1)$ in an open ball $U(a)$. Then there exist functions $g_i(x)$ such that

(1) $f(x) = f(a) + \sum_i (x^i - a^i)\ g_i(x)$

(2) $g_i(x)$ are of class C^{p-1} in U and

(3) $g_i(a) = \left(\frac{\partial f}{\partial x^i}\right)_a$

Here we have written $g_i(x) = g_i(x,a)$.

1.11 Mean Value Theorem for $f: R^n \to R^m$. When we apply Theorem 9 in this case we must consider the component functions separately. For each of these, $f^\alpha(x)$, we get an s_α, but in general the s_α will be distinct. Hence we obtain the result

Theorem 11. Let $f: R^n \to R^m$ be differentiable at every point of a line segment $a + th$ for $0 \leq t \leq 1$. Then there exists a matrix L:

$$L_i^\alpha = \begin{pmatrix} f_1^1 (s_1) & \dots & f_n^1 (s_1) \\ f_1^2 (s_2) & \dots & f_n^2 (s_2) \\ & - - - & \\ f_1^m (s_m) & \dots & f_n^m (s_m) \end{pmatrix}$$

where $f_i^\alpha = \dfrac{\partial f^\alpha}{\partial x^i}$ for $\alpha = 1 \dots m$, $i = 1 \dots n$.

such that $f^\alpha(a + h) - f^\alpha(a) = \sum_i L_i^\alpha h^i$

$$\text{or } f(a + h) - f(a) = Lh$$

The analog of Theorem 10 is now:

Theorem 12. Let $f: R^n \to R^m$ be of class C^p $(p \geq 1)$ in an open convex set U and let x_1 and x_2 be two points in U. Then in $U \times U$, there exist functions $g_i^\alpha (x_1, x_2)$ such that

(1) $f^\alpha(x_2) - f^\alpha(x_1) = \sum_i (x_2^i - x_1^i)\, g_i^\alpha (x_1, x_2)$

(2) $g_i^\alpha(x_1, x_2)$ are of class C^{p-1} in $U \times U$, and

(3) $g_i^\alpha (x, x) = \left(\dfrac{\partial f^\alpha}{\partial x^i}\right)_x$

This may also be written

$$f^\alpha(a + h) = f^\alpha(a) + \sum_i h^i \, g_i^\alpha(h)$$

where we have put $g_i^\alpha(h) = g_i^\alpha(a + h, a)$

so that $g_i^\alpha(h)$ are of class C^{p-1} in U and $g_i^\alpha(0) = \left(\frac{\partial f^\alpha}{\partial x^i}\right)_a$

1.12 Taylor's Theorem. As in the case of one-variable calculus this theorem is an extension and generalization of the Mean Value Theorem. First we prove a lemma that is really an elementary form of Taylor's Theorem.

Lemma. Let $F: R^1 \to R^n$ be defined in $[0,1]$ as a function of class C^p $(p \geq 1)$ with component functions $F^\alpha(t)$. Then for $r \leq p$:

$$F^\alpha(1) = F^\alpha(0) + \frac{dF^\alpha}{dt}(0) + \frac{1}{2!}\frac{d^2F^\alpha}{dt^2}(0) + \dots + \frac{1}{(r-1)!}\frac{d^{r-1}F^\alpha}{dt^{r-1}}(0)$$

$$+ \int_0^1 \frac{(1-t)^{r-1}}{(r-1)!}\frac{d^rF^\alpha}{dt^r}(t)dt$$

Proof: By differentiation we verify that:

$$\frac{d}{dt}\left[F^\alpha(t) + (1-t)\frac{dF^\alpha}{dt} + \dots + \frac{1}{(r-1)!}(1-t)^{r-1}\frac{d^{r-1}F^\alpha}{dt^{r-1}}\right]$$

$$= \frac{1}{(r-1)!}(1-t)^{r-1}\frac{d^rF^\alpha}{dt^r}$$

Then integrating both sides from 0 to 1 we obtain the formula of the lemma.

Corollary: If in addition to the hypotheses of the lemma we assume that $\left|\frac{d^rF^\alpha}{dt^r}\right| \leq M^\alpha$ in $[0,1]$, we find that:

$$|F^\alpha(1) - [F^\alpha(0) + \frac{dF^\alpha}{dt}(0) + \ldots + \frac{1}{(r-1)!}\frac{d^{r-1}F^\alpha}{dt^{r-1}}(0)]|$$

$$\leq M^\alpha |\int_0^1 \frac{(1-t)^{r-1}}{(r-1)!}\,dt| = \frac{M^\alpha}{r!}$$

This is an elementary form of Taylor's Theorem with Lagrange remainder.

Theorem 13. (Taylor's Theorem) Let $f\colon R^n \to R^m$ be of class C^p $(p \geq 1)$ in an open ball $U(a) \subset R^n$, then for $r \leq p$:

$$f^\alpha(a + h) = f^\alpha(a) + \Sigma\left(\frac{\partial f^\alpha}{\partial x^i}\right)_a h^i + \frac{1}{2}\Sigma\left(\frac{\partial^2 f^\alpha}{\partial x^i \partial x^j}\right)_a h^i h^j$$

$$+ \ldots + \frac{1}{(r-1)!}\Sigma\left(\frac{\partial^{r-1} f^\alpha}{\partial x^{i_1}\ldots\partial x^{i_{r-1}}}\right)_a h^{i_1}\ldots h^{i_{r-1}}$$

$$+ \Sigma\int_0^1 \frac{(1-t)^{r-1}}{(r-1)!}\left(\frac{\partial^r f^\alpha}{\partial x^{i_1}\ldots\partial x^{i_r}}\right)_{a+th} h^{i_1}\ldots h^{i_r}\,dt$$

Proof: In the above lemma set $F^\alpha(t) = f^\alpha(a + th)$ and use the formulas:

$$\frac{dF^\alpha}{dt} = \Sigma\left(\frac{\partial f^\alpha}{\partial x^i}\right)_{a+th} h^i$$

$$\frac{d^r F^\alpha}{dt^r} = \Sigma\left(\frac{\partial^r f^\alpha}{\partial x^{i_1}\ldots\partial x^{i_r}}\right)_{a+th} h^{i_1}\ldots h^{i_r}$$

If $\left|\frac{\partial^r f^\alpha}{\partial x^{i_1}\ldots\partial x^{i_r}}\right| \leq M^\alpha$ for $x \in U(a)$, then

$$\left| \Sigma \left(\frac{\partial^r f^\alpha}{\partial x^{i_1} \ldots \partial x^{i_r}} \right)_{a+th} h^{i_1} \ldots h^{i_r} \right| \leq M^\alpha |h|^r$$

So we have the corollary for the Lagrange form of the remainder:

Corollary:

$$|f^\alpha(a+h) - [f^\alpha(a) + \Sigma \left(\frac{\partial f^\alpha}{\partial x^i}\right)_a h^i + \ldots +$$

$$\frac{1}{(r-1)!} \Sigma \left(\frac{\partial^{r-1} f^\alpha}{\partial x^{i_1} \ldots \partial x^{i_{r-1}}} \right)_a h^{i_1} \ldots h^{i_{r-1}}]| \leq \frac{M^\alpha |h|^r}{r!}$$

From Taylor's Theorem we can derive a generalization of Theorem 12:

Theorem 14. Let $f: R^n \to R^m$ be of class C^p $(p \geq 1)$ in an open ball $U(a) \subset R^n$, then for $r \leq p$ there exist functions $g^\alpha_{i_1 \ldots i_r}(h)$ such that for $h \in U(a)$:

(1) $$f^\alpha(a+h) = f^\alpha(a) + \Sigma \left(\frac{\partial f^\alpha}{\partial x^i}\right)_a h^i + \ldots +$$

$$\frac{1}{(r-1)!} \Sigma \left(\frac{\partial^{r-1} f^\alpha}{\partial x^{i_1} \ldots \partial x^{i_{r-1}}} \right)_a h^{i_1} \ldots h^{i_{r-1}}$$

$$+ \frac{1}{r!} \Sigma \, g^\alpha_{i_1 \ldots i_r}(h) \, h^{i_1} \ldots h^{i_r}$$

(2) $g^\alpha_{i_1 \ldots i_r}(h)$ are of class C^{p-r} in $U(a)$

(3) $$g^\alpha_{i_1 \ldots i_r}(0) = \left(\frac{\partial^r f^\alpha}{\partial x^{i_1} \ldots \partial x^{i_r}} \right)_a$$

Proof:

Let $g^{\alpha}_{i_1 \ldots i_r}(h) = r\int_0^1 (1-t)^{r-1} \left(\frac{\partial^r f^{\alpha}}{\partial x^{i_1} \ldots \partial x^{i_r}} \right)_{a+th} dt$

Then $g^{\alpha}_{i_1 \ldots i_r}(h)$ have the properties stated in the theorem.

1.13 Integration. Let $f\colon R^n \to R^m$. Then $I = \int_D f$, where D is a region in R^n, can be interpreted by means of the coordinate functions of f as the element I of R^m such that

$$I^{\alpha} = \int_D f^{\alpha}$$

Theorem 15. $\left|\int_D f\right| \leq \int_D |f|$ for f continuous in D.

Proof. Let f be a step function defined in D. That is, let D be partitioned into a collection of subsets D_i such that $f|D_i = c_i$. Let the measure of D_i be $m(D_i)$.

Then

$$\int_D f = \sum_i c_i \, m(D_i) \quad \text{and} \quad \int_D |f| = \sum_i |c_i| m(D_i)$$

and $\left|\int_D f\right| \leq \sum_i |c_i| \, m(D_i)$ (by the triangle inequality)

$$\leq \int_D |f|.$$

Since the integral of a continuous function is the limit of a sequence of integrals of step functions, the inequality holds when f is continuous.

Exercises

1. Prove the converse of Theorem 2.

2. Let $f\colon R^n \to R^m$.

   Prove: $\lim_{x \to a} f(x) = y$ iff $\lim_{x \to a} f^\alpha(x) = \pi^\alpha(y)$; $\alpha = 1 \ldots m$

3. Prove: If $f\colon R^n \to R^m$ is constant, then $df_a = 0$ (the zero transformation).

4. Prove: If $f\colon R^n \to R^m$ is linear, then $df_a = f$ for all $\underline{a}$.

5. Prove the result stated in example 1.

6. Prove the result stated in example 2.

7. Prove the result stated in example 3.

8. Let $f\colon R^n \to R^m$ be differentiable in U. Consider a change of basis in R^m so that the component functions $\bar{f}^\beta(x)$ in the new basis are given by

$$\bar{f}^\beta(x) = \sum_\alpha A^\beta_\alpha f^\alpha(x)$$

   Also consider a change of basis in R^n so that $x^i = \sum_j B^i_j \bar{x}^j$

   Prove that

$$\frac{\partial \bar{f}^\beta}{\partial \bar{x}^i} = \sum_{\alpha,i} A^\beta_\alpha \frac{\partial f^\alpha}{\partial x^i} B^i_j$$

   Hence show that the definition of a C^1 function is independent of the bases chosen in the definition.

9. Prove that the function below is differentiable at $(0,0)$ even though its partial derivatives (which do exist) are not continuous at $(0,0)$.

$$f(x,y) = \begin{cases} (x^2 + y^2) \sin \dfrac{1}{\sqrt{x^2 + y^2}} & (x,y) \neq (0,0) \\ 0 & (x,y) = (0,0) \end{cases}$$

10. Prove: If $f: R^n \to R^m$ is linear, then the affine function $A(x - a)$ of 1.7 equals $f(x)$ for all $\underline{a}$. Hence in this case the approximation is exact.

11. For what directions does

$$f(x,y) = \begin{cases} \dfrac{2xy}{x^2 + y^2} & (x,y) \neq (0,0) \\ 1 & (x,y) = (0,0) \end{cases}$$

have directional derivatives?

12. Suppose that we are given a fixed (but unknown) vector $\underline{v}$ at a point $\underline{a}$ in R^n and that we are given $\dfrac{df^i}{dv}$ at $\underline{a}$ for each of n known functions $f^i(x)$ such that $\det\left(\dfrac{\partial f^i}{\partial x^j}\right) \neq 0$ at $\underline{a}$. Prove that this information allows us to compute the components of $\underline{v}$.

13. Prove that $\dfrac{|L(h)|}{|h|}$ is bounded where L is a linear transformation.

14. Prove: If $f: R^n \to R^m$ and $g: R^1 \to R^1$ are differentiable at every point of the line segment $a + th$ for $0 \le t \le 1$, and

$$|df_{a+th}| \le \frac{dg}{dt}(a + th) \quad \text{for all } t \text{ in } [0,1],$$

then $|f(a + h) - f(a)| \le g(a + h) - g(a)$

Hint: $f(a + h) - f(a) = \int_0^1 \frac{df}{dt}(a + th)dt$

$$= \int_0^1 df_{a + th}(h)dt$$

15. Prove: If $f: R^n \to R^m$ is differentiable at every point of a line segment $a + th$ for $0 \le t \le 1$, then

$$|f(a + h) - f(a)| \le |h| \sup |df_\xi|$$

where the sup is taken over all ξ in the segment.

16. Prove: If $f: R^n \to R^m$ is differentiable at every point of a line segment $a + th$ for $0 \le t \le 1$, and if x_0 is an arbitrary point on the line segment, then

$$|f(a + h) - f(a) - df_{x_0}(h)| \le |h| \sup |df_\xi - df_{x_0}|$$

where the sup is taken over all ξ in the segment.

Hint: Apply exercise 15 to $g(x) = f(x) - df_{x_0}(x)$

17. Prove: If $f: R^n \to R^n$ is of class C^1 in an open convex set $U(a)$, and if df_a is nonsingular, then there is an open set $V(a) \subset U(a)$ such that for x and y in $V(a)$

$$f(x) = f(y) \text{ iff } x = y.$$

18. Prove: If $f: R^n \to R^m$ is differentiable in an open convex set U, and if there is a $C \geq 0$ such that $|df_a| \leq C$ for every $\underline{a}$ in U, then for every x and y in U

$$|f(x) - f(y)| \leq C|x - y|$$

19. Prove: If $f: R^n \to R^m$ is differentiable in an open piecewise linearly connected set U and if $df_a = 0$ for every $\underline{a}$ in U, then f is constant in U. (Assume that every two points x and y in U can be joined by a broken line segment in U.)

Chapter 2

Existence Theorems for Differential Equations

2.1 Introduction. Consider $f(x,t): R^n \times R^1 = R^{n+1} \to R^n$ which is a $C^p (p \geq 0)$ function (or an analytic function) defined on $V \times U \subset R^{n+1}$ where

$V = \{x: |x - x_o| \leq b\}$ is a closed ball in R^n

$U = \{t: |t - t_o| \leq a\}$ is a closed interval in R^1.

We shall discuss the differential equation

(1) $$\frac{dx}{dt} = f(x,t)$$

A solution of (1) is a function $g(t): R^1 \to R^n$ such that

$$\frac{dg(t)}{dt} = f[g(t),t]$$

for all t in a stated subinterval of U. A solution satisfies the initial condition (x_o, t_o) provided that $g(t_o) = x_o$.

Our purpose is to establish the existence and uniqueness of such solutions under stated conditions.

2.2 The Basic Theorem.

Theorem 1. If f is of class C^p $(p \geq 1)$ (or analytic), there is a subset $\overline{U}$ of U containing t_o within which there is a unique solution of (1) of class C^{p+1} (or analytic) satisfying the initial condition (x_o, t_o).

If f is merely continuous, the above conclusion holds provided that f satisfies the Lipschitz Condition:

(2) $|f(x_2,t) - f(x_1,t)| \leq K|x_2 - x_1|$ for some $K > 0$

for all x_1 and x_2 in V and all t in U.

Note that if f is differentiable with bounded partial derivatives with respect to x, or if f is of class C^1 or higher, this Lipschitz Condition is automatically satisfied. (See exercise 18, Chapter 1).

There are many proofs of this theorem, and we present two to illustrate the various methods.

2.3 <u>The Picard Proof by Successive Approximations</u>. This is a slight modification of the standard Picard proof so that it will be closely related to the "contraction map" proof which follows.

Before giving the proof, we illustrate the method by an elementary example. Consider the differential equation in $R^1 \times R^1$; $\frac{dx}{dt} = x$ with initial condition; $x = 1$, $t = 0$, whose solution is well known to be $x = e^t$. This differential equation is equivalent to the integral equation

$$g(t) = 1 + \int_0^t g(u)du$$

for, by differentiation, we obtain $\frac{dg}{dt} = g(t)$ and $g(0) = 1$.

The method of approximation proceeds in the following steps:

$$\text{Put } g_0(t) = 1$$

$$g_1(t) = 1 + \int_0^t g_0(u)du$$

$$= 1 + \int_0^t 1\, du = 1 + t$$

$$g_2(t) = 1 + \int_0^t g_1(u)du$$

$$= 1 + \int_0^t 1 + u\, du = 1 + t + \frac{t^2}{2}$$

By induction, suppose that

$$g_n(t) = 1 + t + \dots + \frac{t^n}{n!}$$

Then

$$g_{n+1}(t) = 1 + \int_0^t 1 + u + \dots + \frac{u^n}{n!}\, du$$

$$= 1 + t + \frac{t^2}{2!} + \dots + \frac{t^{n+1}}{(n+1)!}$$

Consider the sequence: $g_0, g_1, \dots, g_n, \dots$ This converges uniformly to e^t. We check by showing that $g(t) = e^t$ is, indeed, the desired solution.

<u>General Case</u>. We assume V and U as in (2.1), that f satisfies the Lipschitz Condition (2) and that the initial condition is (x_0, t_0). Further let M be the maximum value of $|f|$ in $V \times U$ and h be strictly less than the least of:

a, b/M, and $1/K$ (where K is the Lipschitz constant)

and let $\overline{U}$ be the interval $|t-t_0| \leq h$. Hereafter we assume that $t \in \overline{U}$. The equivalent integral equation is:

(3) $$g(t) = x_0 + \int_{t_0}^t f[g(u),u]\, du$$

The approximation then proceeds as follows:

$$g_0(t) = x_0$$

$$g_1(t) = x_0 + \int_{t_0}^t f[g_0(u),u]\, du$$

$$g_2(t) = x_0 + \int_{t_0}^t f[g_1(u),u]\, du$$

$$g_{n+1}(t) = x_0 + \int_{t_0}^t f[g_n(u),u]\, du$$

To prove the convergence of this sequence of functions to a solution of (1), we introduce a series of lemmas.

<u>Lemma 1</u>. $|g_n(t) - x_o| \leq b$ for $|t - t_o| \leq h$

This shows that $g_n(t)$ stays in the set V. Proof by induction:

(a) $|g_1(t) - x_o| \leq \int_{t_o}^{t} |f[g_o(u),u]|\, du$

$$\leq M|t - t_o| \leq Mh \leq b$$

(b) Suppose that $|g_n(t) - x_o| \leq b$ for $|t - t_o| \leq h$

Then $|f[g_n(t),t]| \leq M$

So $|g_{n+1}(t) - x_o| \leq \int_{t_o}^{t} |f[g_n(u),u]|\, du$

$$\leq M|t - t_o| \leq Mh \leq b$$

<u>Lemma 2</u>. The series

$$x_o + \Sigma[g_n(t) - g_{n-1}(t)]$$

converges uniformly in $|t - t_o| \leq h$.

<u>Proof</u>. From the Weierstrass M-test this follows if

$$|g_n(t) - g_{n-1}(t)| \leq \frac{M}{K}(Kh)^n \quad \text{where} \quad Kh < 1.$$

Then by induction:

(a) $|g_1(t) - g_o(t)| \leq Mh = \frac{M}{K}(Kh)$

as in (a) of Lemma 1.

(b) Suppose that

$$|g_n(t) - g_{n-1}(t)| \leq \frac{M}{K}(Kh)^n$$

Then

$$|g_{n+1}(t) - g_n(t)| \leq \int_{t_0}^{t} |f[g_n(u),u] - f[g_{n-1}(u),u]|du$$

$$\leq \int_{t_0}^{t} K|g_n(u) - g_{n-1}(u)|du$$

$$\leq K \frac{M}{K} (Kh)^n |t - t_0|$$

$$\leq \frac{M}{K} (Kh)^{n+1}$$

So lemma 2 is proved.

The nth partial sum, S_n, of this series is clearly $g_n(t)$. Thus $\lim_{n\to\infty} g_n(t)$ exists and is continuous. We call this limit $g(t)$.

<u>Lemma 3</u>. The function $g(t)$ is a solution of the integral equation (3) and hence of the differential equation (1). Moreover, it satisfies the given initial conditions.

<u>Proof</u>. Let $f_n(t) = f[g_n(t),t]$

Then the series

$$\Sigma[(f_n(t) - f_{n-1}(t)]$$

is uniformly convergent, for

$$|f_n(t) - f_{n-1}(t)| = |f[g_n(t),t] - f[g_{n-1}(t),t]|$$

$$\leq K |g_n(t) - g_{n-1}(t)|$$

$$\leq M (Kh)^n$$

So the sequence $\{f_n(t)\}$ is uniformly convergent to $f[g(t),t]$.

Finally

$$g(t) = \lim_{n\to\infty} g_n(t) = x_0 + \lim_{n\to\infty} \int_{t_0}^{t} f[g_{n-1}(u),u]du$$

$$= x_0 + \lim_{n\to\infty} \int_{t_0}^{t} f_{n-1}(u)du$$

$$= x_0 + \int_{t_0}^{t} \lim_{n\to\infty} f_{n-1}(u)du$$

$$= x_0 + \int_{t_0}^{t} f[g(u),u]du$$

This is what is required to prove the lemma.

Unicity of Solution. Suppose that there is another solution $G(t)$ of equation (3) with the same initial condition. Then

$$G(t) = x_0 + \int_{t_0}^{t} f[G(u),u]du$$

The fact that $G(t) = g(t)$ follows from Lemma 4.

Lemma 4. $|G(t) - g_n(t)| < b(Kh)^n$

Proof (by induction):

(a) $$|G(t) - g_1(t)| \leq \int_{t_0}^{t} |f[G(u),u] - f[g_o(u),u]|du$$

$$\leq \int_{t_0}^{t} K|G(u) - g_o(u)|du$$

$$\leq \int_{t_0}^{t} K\ b\ du \leq b(Kh)$$

(b) If $|G(t) - g_n(t)| \leq b(Kh)^n$, then

$$|G(t) - g_{n+1}(t)| \leq \int_{t_0}^{t} |f[G(u),u] - f[g_n(u), u]|du$$

$$\leq \int_{t_0}^{t} K|G(u) - g_n(u)|du$$

$$\leq \int_{t_0}^{t} Kb\,(Kh)^n\,du$$

$$\leq b(Kh)^{n+1}$$

Therefore, since $Kh < 1$, $\lim_{n\to\infty} g_n(t) = G(t)$. So $g(t) = G(t)$.

Estimate of Error at n^{th} step.

$$g(t) - g_n(t) = \sum_{k=n+1}^{\infty} g_k(t) - g_{k-1}(t)$$

So $$|g(t) - g_n(t)| \leq \sum_{k=n+1}^{\infty} |g_k(t) - g_{k-1}(t)|$$

$$\leq \frac{M}{K} \sum_{n+1}^{\infty} (Kh)^k \qquad \text{[from Lemma 2]}$$

$$\leq \frac{M}{K} (Kh)^{n+1} \sum_{j=0}^{\infty} (Kh)^j$$

$$\leq \frac{M}{K} (Kh)^{n+1} \cdot \frac{1}{1-Kh}$$

2.4 Contraction Map Proof. Many modern mathematicians, especially functional analysts, prefer an apparently different approach to this subject. Here we give an introductory treatment based on the so-called Contraction Map. Actually this is nothing more than an abstract repackaging of the Picard proof. The domain of discussion is a Banach Space.

<u>Definition</u>. A Banach space is a normed topological vector space (over the reals) whose topology is given by the norm and which is complete. (Example R^n).

<u>Lemma 5. Contraction Map</u>. Let B be a Banach space with the norm of an element $p \in B$ denoted by $\| p \|$. Let T be a transformation on the set W of all p satisfying $\| p \| \leq b$ (for $b > 0$) which is such that Tp is in W and which satisfies the Lipschitz condition.

$$\| Tp - Tq \| \leq L \| p - q \| \qquad \text{where} \quad L < 1$$

for all p and q in W. Then there is a unique element of W, r, such that $Tr = r$; that is, r is a fixed point of T.

(a) <u>Unicity</u> is trivial; for if r_1 and r_2 are distinct fixed points, then $\| Tr_1 - Tr_2 \| = \| r_1 - r_2 \|$, which is impossible because of Lipschitz.

(b) <u>Existence</u>. In fact $r = \lim\limits_{n\to\infty} T^n p$ where p is any element of W. For convenience choose $p = 0$.

From the Lipschitz condition

$$\begin{aligned} \| T^{n+1}(0) - T^n(0) \| &\leq L \, \| T^n(0) - T^{n-1}(0) \| \\ &\leq L^2 \| T^{n-1}(0) - T^{n-2}(0) \| \\ &\quad - - - \\ &\leq L^n \| T(0) - 0 \| \leq L^n b \end{aligned}$$

Hence

$$\| T^{n+m}(0) - T^n(0) \| \leq (L^n + L^{n+1} + \ldots + L^{n+m-1}) b$$

But $L^n + \dots + L^{n+m-1}$ is the Cauchy difference of a convergent geometric series and hence can be made arbitrarily small by choosing n and m large enough. Hence $T^n(0)$ is a Cauchy sequence, which must converge since B is complete.

Application to Differential Equations.

Consider equation (1) of section 2.1 where without loss of generality we assume $x_0 = 0$. Let B be the Banach space of continuous functions $p(t)\colon R^1 \to R^n$ defined on $|t - t_0| \le h$ and with $|p(t)| \le b$. The norm $\|p(t)\|$ is chosen to be $\max_t |p(t)| \le b$.

Define $Tp = \int_{t_0}^{t} f[p(u),u]du$

Then $\|Tp\| \le b$ by Lemma 1.

Also

$$\|Tp - Tq\| \le \int_{t_0}^{t} |f[p(u),u] - f[q(u),u]|du$$

$$\le Kh\ \|p - q\| \quad \text{where} \quad Kh < 1$$

So the Lipschitz condition of the Banach space is satisfied with $L = Kh$.

Then apply Lemma 5. The result is that there is a unique fixed point $r(t)$ of B satisfying

$$r(t) = x_0 + \int_{t_0}^{t} f[r(u),u]du$$

and hence the existence theorem follows.

2.5 <u>Refinements</u>.

(1) In Sections 2.3 and 2.4 it was assumed that f is of class C^0 and that the Lipschitz condition was satisfied. If only C^0 is assumed, a different proof shows that a solution exists, but it may well not be unique.

<u>Example</u>. The equation $dx/dt = \sqrt{|x|}$ has more than one solution with initial values $(0,0)$. Note that the Lipschitz Condition is not satisfied. (See exercise 2.)

(2) Our requirement that t be in $|t - t_0| \leq h$ is unduly restrictive. Theorems exist which permit this interval to be enlarged.

(3) Our proof shows that if f is of class C^0 then g is of class C^1. But we must still show that if f is of class C^p $(p \geq 1)$, g is of class C^{p+1}.

We know that

$$\frac{dg}{dt} = f\,[g(t),t]$$

Now if f is of class C^1, then the right side is C^1 in t. So the left side is C^1 and $g(t)$ is C^2. By induction the result follows.

(4) The argument in (3) extends to C^∞, but the analytic case must be handled separately. This involves going to the complex case, using the theory of analytic functions of a complex variable, and then coming back down to the reals.

2.6 <u>Approximate Solutions</u>.

<u>Definition</u>. A C^1 function $a(t)$ defined in $|t - t_0| \leq h$ is an ε-approximate solution of the differential equation (1) iff:

$$\left|\frac{da}{dt} - f[a(t),t]\right| \leq \varepsilon \quad \text{for all} \quad t \quad \text{in} \quad |t - t_o| \leq h$$

An important theorem concerning approximate solutions is as follows:

<u>Theorem 2</u>. If $f(x,t)$ is as in Theorem 1, if $g(t)$ is a solution of (1) and $a(t)$ is an ε-approximate solution of (1) with the same initial value, then

$$|g(t) - a(t)| \leq \frac{\varepsilon h}{1-Kh}$$

<u>Proof</u>. By hypothesis

$$|a(t) - x_o - \int_{t_o}^{t} f[a(u),u]du| \leq \varepsilon h$$

$$|g(t) - x_o - \int_{t_o}^{t} f[g(u),u]du| = 0$$

Hence $$|g(t) - a(t) - \int_{t_o}^{t} f[g(u),u] - f[a(u),u]du| \leq \varepsilon h$$

Let $\Delta(t) = |g(t) - a(t)|$

Then we have

$$\Delta(t) \leq \int_{t_o}^{t} |f[g(u),u] - f[a(u),u]|du + \varepsilon h$$

$$\leq K \int_{t_o}^{t} \Delta(u)du + \varepsilon h$$

$$\max \Delta(t) \leq Kh \max \Delta(t) + \varepsilon h$$

or $\max \Delta(t) \leq \dfrac{\varepsilon h}{1-Kh}$. Hence $\Delta t \leq \dfrac{\varepsilon h}{1-Kh}$

2.7 Linear Equations. A linear system of differential equations has the form:

$$\frac{dx}{dt} = P_t(x) \quad \text{where } P_t \text{ is a linear transformation whose}$$

matrix relative to a basis has elements that are continuous in U. The corresponding theorem is:

Theorem 3. There exists a unique solution of a linear equation in $\overline{U}$ with given initial conditions (x_0, t_0).

Proof. The right hand side is continuous, and all that is required is to show that Lipschitz is satisfied.

Let x_1 and x_2 be any two elements of V.

Then

$$|P_t(x_1 - x_2)| \leq |P_t|\ |x_1 - x_2| \leq M|x_1 - x_2|$$

where M is the maximum of $|P_t|$ in U.

Remark. This theorem is of great value in proving that a given Function F is zero (not an easy thing to do otherwise). First show that F satisfies a linear differential equation with initial value $x_0 = 0$ at $t = t_0$. There is an obvious solution of the differential equation with this initial value, namely $g(t) \equiv 0$ and this solution is unique. Hence F must be this solution and so $F \equiv 0$.

2.8 Dependence of Solutions on Initial Values and Parameters.

The function g(t) which satisfies

$$g(t) = x_0 + \int_{t_0}^{t} f[g(u), u]du$$

clearly depends on x_0. So we may write it as $g(x_0,t)$. The following theorem is true:

<u>Theorem 4</u>. If $f(x,t)$ is as in Theorem 1, then the solution $g(x_0,t)$ is of class C^p (or analytic) in x_0.

The proof requires these steps:

(a) <u>Continuity</u>. This follows from Lemma 6.

<u>Lemma 6</u>.

$$|g_n(x_0 + s,t) - g_n(x_0,t)| \leq |s| [1 + (Kh) + \dots + (Kh)^n]$$

The proof is an easy induction using the Lipschitz Condition.

In this lemma take the limit as $n \to \infty$. The result is

$$|g(x_0 + s,t) - g(x_0,t)| \leq \frac{|s|}{1-Kh}$$

Hence $g(x_0,t)$ is continuous in x_0.

(b) <u>Differentiability</u>. First we prove Lemma 7:

<u>Lemma 7</u>. If $f(x,t)$ is of class C^1, then $g(x_0,t)$ is of class C^1 in x_0.

<u>Proof</u>: If we knew that $g(x_0,t)$ was of class C^1, then since

$$(3') \quad g^i(x_0,t) = x_0^i + \int_{t_0}^t f^i[g(u),u]du$$

it would follow that

$$\frac{\partial g^i(x_0,t)}{\partial x_0^j} = \delta_j^i + \int_{t_0}^t \sum_k \frac{\partial f^i}{\partial x^k}[g(u),u]\frac{\partial g^k}{\partial x_0^j}(x_0,u)du$$

Hence $\dfrac{\partial g^i(x_o,t)}{\partial x_o^j}$ would satisfy the linear differential equation:

(4) $$\frac{d}{dt}(y_j^i) = \sum_k \frac{\partial f^i}{\partial x^k} [g(x_o,t)t]\, y_j^k$$

with initial values δ_j^i. Our problem is to make this derivation legitimate.

Define $G_j^i(x_o,s,t) = g^i(x_o + se_j,t) - g^i(x_o,t)$

Then by Lemma 6, G_j^i are uniformly continuous in s. Since $g^i(x_o,t)$ and $g^i(x_o + se_j,t)$ are solutions of (3'), it follows that:

$$\frac{dG_j^i}{dt} = f^i[g(x_o + se_j,t),t] - f^i[g(x_o,t),t]$$

Since f^i are of class C^1, we obtain from the Mean Value Theorem that there exist uniformly continuous functions h_j^i such that:

$$\frac{dG_j^i}{dt} = \sum_k h_k^i [g(x_o + se_j,t),g(x_o,t)]\, G_j^k (x_o,s,t)$$

Using Theorem 12, Chapter 1, we may write

$$h_k^i[g(x_o + se_j,t),g(x_o,t)] = \frac{\partial f^i}{\partial x^k} [g(x_o,t)] + E_k^i(x_o,s,t)$$

where $E_k^i(x_o,s,t) \to 0$ uniformly as $s \to 0$.

Now define

$$P^i_j(x_o,s,t) = \frac{G^i_j(x_o,s,t)}{s}$$

Then $P^i_j(x_o,s,t_o) = \dfrac{x^i_o + s\delta^i_j - x^i_o}{s} = \delta^i_j$; $|P^i_j(x_o,s,t_o)| \leq \dfrac{1}{1-Kh}$

and $$\frac{d}{dt}(P^i_j) = \sum_k \frac{\partial f^i}{\partial x^k}[g(x_o,t)]P^k_j + \sum_k E^i_k P^k_j$$

From the above analysis it follows that

$|\sum_k E^i_k P^k_j| \to 0$ uniformly as $s \to 0$; and in particular

$|\sum_k E^i_k P^k_j| < \varepsilon$ if $|s| < \delta(\varepsilon)$. Therefore, $P^i_j(x_o,s,t)$ are ε-approximate solutions of the differential equation (4) with initial values δ^i_j.

From Theorem 2, section 2.6, we may now conclude that

$$|Q^i_j(x_o,t) - P^i_j(x_o,s,t)| \leq \varepsilon A$$

for some constant A where $Q^i_j(x_o,t)$ are solutions of (4) with initial values δ^i_j. Therefore:

$$\lim_{s\to 0} P^i_j(x_o,s,t) = Q^i_j(x_o,t)$$

Hence $\dfrac{\partial g^i(x_o,t)}{\partial x^j_o}$ exist and are continuous.

(c) Higher Differentiability. This requires the use of Theorem 5 which follows.

At this point we shift to the following problem. Suppose that $f(x,t)$ also involves a parameter, λ, so that we have $f(x,\lambda,t)$. The corresponding solution can then be written $g(\lambda,t)$. We then have the result:

Theorem 5. If $f(x,\lambda,t)$ is C^p, then $g(\lambda,t)$ is also C^p.

Proof for C^1 case. Write the differential equation in terms of component functions:

$$\frac{dx^i}{dt} = f^i(x,\lambda,t) \qquad i = 1\ldots n$$

Then consider the system

$$\begin{cases} \dfrac{dx^i}{dt} = f^i(x,y,t) & \text{where } y \text{ is in } R^1. \\[2ex] \dfrac{dy}{dt} = 0 \end{cases}$$

with initial conditions at t_o: $x = x_o, y = \lambda$. This has a unique solution:

$$x^i = g^i(x_o,\lambda,t); \quad y = \lambda$$

By Lemma 7, $g^i(x_o,\lambda,t)$ are C^1 in λ.

Now return to the (c) part of Theorem 4. We know that $\dfrac{\partial g^i(x_o,t)}{\partial x_o^j}$ satisfy the linear differential equation (4); where now each component of x_o can be considered to be a parameter.

If we assume that f is C^2, the right side is C^1 in x_o. Hence y^i_j are C^1, or $g^i(x_o,t)$ are C^2.

Then we prove Theorem 5 for the C^2 case, then Theorem 4 for the C^3 case, and so on by induction. As before the analytic case must be handled by methods of complex variable theory.

2.9 Partial Differential Equations - First Problem. In this section we consider a classical problem: When is a vector field in R^n a gradient? We are given the vector field $v_i(x)$ $i = 1...n$, $x \in R^n$; which is another way of thinking of a function $v: R^n \to R^n$. The question is whether there is a function $F: R^n \to R^1$ such that $\partial F/\partial x^i = v_i$. If so, the vector field v_i is called the gradient of F.

As a theorem in partial differential equations we recast the problem this way:

To solve: $$\frac{\partial F}{\partial x^i} = v_i(x)$$

with the initial condition $F(x_o) = F_o$. In contrast to the theory of ordinary differential equations, a solution need not exist. Indeed certain conditions, called integrability conditions must be satisfied in order that there be a solution. The relevant theorem in this case is Theorem 6.

Theorem 6. Let $v_i(x)$ be C^1 functions defined in an open simply connected region D of R^n. Then there exists a unique solution of the differential equation

$$\frac{\partial F}{\partial x^i} = v_i(x) \quad \text{such that} \quad F(x_o) = F_o$$

iff the following integrability conditions are satisfied in D:

$$\frac{\partial v_i}{\partial x^j} = \frac{\partial v_j}{\partial x^i}$$

Remark. This result may well be false if D is not simply connected.

Necessity is immediate; for if there is a solution

$$\frac{\partial v_i}{\partial x^j} = \frac{\partial^2 F}{\partial x^j \partial x^i} = \frac{\partial^2 F}{\partial x^i \partial x^j} = \frac{\partial v_j}{\partial x^i}$$

To prove sufficiency we construct F by means of line integration. Let P have coordinates x_o^i and be the initial point in D. Choose any other point Q in D and join P and Q with a C^1 curve $x^i = f^i(t)$ where $f^i(t)$ are C^1 functions such that $f^i(0)$ is P and $f^i(1)$ is Q. Then consider

$$F(Q) = F_o + \int_0^1 \sum_i v_i[x(t)] \frac{dx^i}{dt} dt$$

Since Q is arbitrary, this results in a function F(x) defined in D such that at P, $F(P) = F_o$.

It still remains to prove that F(x) satisfies the differential equation. To get at this we must first show that F(Q) depends only on Q and not on the chosen curve joining P and Q; in other words we require that the line integration is independent of the path.

To handle this problem we must discuss the concept of homotopic curves. Let C_1 and C_2 be two C^1 curves in D joining P and Q. Let these be given by the equations:

$$C_1: \quad x^i = f_1^i(t) \quad \text{with} \quad P = f_1^i(0) \quad \text{and} \quad Q = f_1^i(1)$$

$$C_2: \quad x^i = f_2^i(t) \quad \text{with} \quad P = f_2^i(0) \quad \text{and} \quad Q = f_2^i(1)$$

Definition. Let S be the unit square in the (ε, t) plane. Then C_1 and C_2 are C^{1+} homotopic in D iff there is a C^{1+} map $S \to D$ given by

$$x^i = \phi^i(\varepsilon, t)$$

where C^{1+} means C^1 and that $\frac{\partial^2 \phi^i}{\partial \varepsilon \partial t}$ and $\frac{\partial^2 \phi^i}{\partial t \partial \varepsilon}$ exist and are continuous and where:

(1) $\phi^i(\varepsilon, 0) = P$ for all ε

(2) $\phi^i(\varepsilon, 1) = Q$ for all ε

(3) $\phi^i(0, t) = f_1^i(t)$

(4) $\phi^i(1, t) = f_2^i(t)$

This implies that $\left(\frac{\partial \phi^i}{\partial \varepsilon}\right)_{t=0} = 0; \left(\frac{\partial \phi^i}{\partial \varepsilon}\right)_{t=1} = 0.$

Remark. Recall that D is simply connected if for each pair of points P and Q in D every continuous curve joining P and Q is continuously homotopic to every other such curve. Standard approximation theorems show that when D is simply connected, every pair of C^1 curves joining P and Q is C^{1+} homotopic.

Now to prove that the given integrability conditions assure us that the line integral above is independent of the path, we define $F(\varepsilon, t)$ on the square S by the integral

$$F(\varepsilon,t) = F_0 + \int_0^t \sum_i v_i[\phi(\varepsilon,u)] \frac{\partial\phi^i}{\partial u}(\varepsilon,u)\,du$$

Note that as a result of this definition $\left(\frac{\partial F}{\partial \varepsilon}\right)_{t=0} = 0$

Then it follows that

$$(5) \quad \frac{\partial F(\varepsilon,t)}{\partial t} = \sum_i v_i \frac{\partial\phi^i}{\partial t}$$

We may also write

$$(6) \quad \frac{\partial F(\varepsilon,t)}{\partial \varepsilon} = \sum_i v_i \frac{\partial\phi^i}{\partial \varepsilon} + \sigma(\varepsilon,t)$$

where $\sigma(\varepsilon,t)$ is so defined.

Then from (5)

$$\frac{\partial^2 F}{\partial\varepsilon\partial t} = \sum_{i,j} \frac{\partial v_i}{\partial x^j} \frac{\partial\phi^j}{\partial\varepsilon} \frac{\partial\phi^i}{\partial t} + v_i \frac{\partial^2\phi^i}{\partial\varepsilon\partial t}$$

and from (6)

$$\frac{\partial^2 F}{\partial t\partial\varepsilon} = \sum_{i,j} \frac{\partial v_i}{\partial x^j} \frac{\partial\phi^j}{\partial t} \frac{\partial\phi^i}{\partial\varepsilon} + v_i \frac{\partial^2\phi^i}{\partial t\partial\varepsilon} + \frac{\partial\sigma}{\partial t}$$

Then by subtraction we find:

$$0 = \sum_{i,j} \left(\frac{\partial v_i}{\partial x^j} - \frac{\partial v_j}{\partial x^i}\right) \frac{\partial\phi^j}{\partial t} \frac{\partial\phi^i}{\partial\varepsilon} + \frac{\partial\sigma}{\partial t}$$

By hypothesis we know that the integrability condition

$$\frac{\partial v_i}{\partial x^j} - \frac{\partial v_j}{\partial x^i} = 0$$

is satisfied. So $\partial\sigma/\partial t = 0$.

Now consider (6) at $t = 0$. We get

$$0 = 0 + \sigma(\varepsilon,0)$$

But $\frac{\partial\sigma}{\partial t} = 0$, and so $\sigma(\varepsilon,t) = 0$

Finally consider (6) at $t = 1$. We get

$$\left(\frac{\partial F}{\partial \varepsilon}\right)_{t=1} = 0$$

and therefore the line integral above is independent of the path.

Finally we use this independence of the path to prove that F satisfies the given differential equation. To do so let Q have coordinates $\underline{a}$. Connect P and Q with a C^1 curve which is the union of PR and RQ where R has coordinates $a - e_j h$. Here RQ is a line segment parallel to the x_j-axis. The equation of RQ can be written $x = f(t)$ where

$$f = (a - e_j h)(1-t) + at = a + e_j h(t - 1)$$

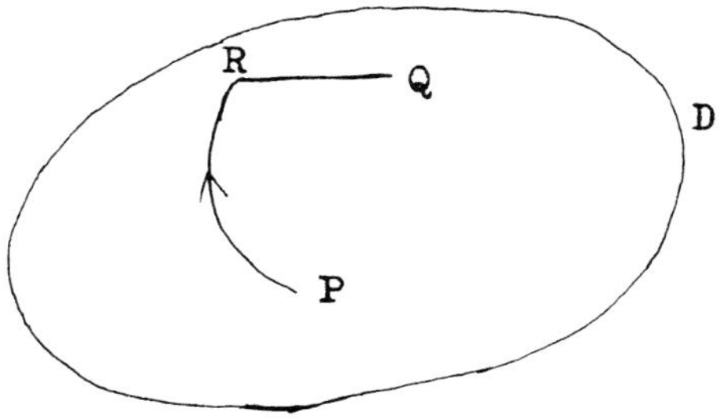

Thus $t = 0$ corresponds to R and $t = 1$ corresponds to Q. Also

$$\frac{df}{dt} = e_j h$$

Then along RQ

$$F[a + e_j h(t - 1)] = F(R) + \int_0^t \{v_j[a + e_j h(u - 1)]\}h\, du$$

and

$$\frac{\partial F}{\partial x^j} = \frac{1}{h}\frac{dF}{dt} = v_j[a + e_j h(t - 1)]$$

So at Q where $t = 1$:

$$\left(\frac{\partial F}{\partial x^j}\right)_Q = v_j(a) = v_j(Q)$$

This completes the proof except for the unicity of F. If there is another solution $G(x)$, then

$$\frac{\partial F}{\partial x^i} - \frac{\partial G}{\partial x^i} = 0$$

Hence $\qquad F(x) = G(x) + C$

But $\qquad F(P) = G(P)$, so $C = 0$ and $F(x) = G(x)$.

Note that in this section we have not used the existence theory for ordinary differential equations.

2.10 The Problem of Section 2.9 When D Is Not Simply Connected

In this section we no longer assume that D is simply connected, and to keep matters easy, restrict ourselves to the plane. To begin with, let D be the plane with the origin deleted. Then the curves C_1 and C_2 joining P and Q in figure 1 are not homotopic.

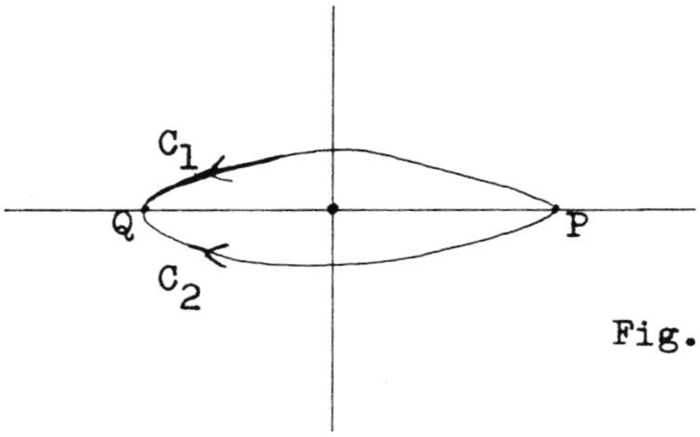

Fig. 1

As a special case suppose that C_1 and C_2 are as shown in figure 2.

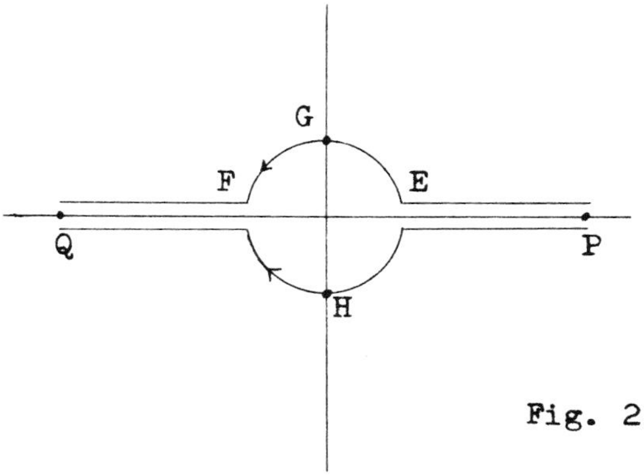

Fig. 2

Observe that $C_1 = PE + EG + GF + FQ$; $C_2 = PE + EH + HF + FQ$

Hence

$$\int_{C_1} \sum_i v_i dx^i - \int_{C_2} \sum_i v_i dx^i = \int_{\Sigma} \sum_i v_i dx^i$$

where $\Sigma = EG + GF - EH - HF$ is a simple closed curve about the origin. Hence for independence of the path, we require that

$$\int_{\Sigma} \sum_i v_i dx^i = 0.$$

It is a true fact (which we shall not prove) that any curve joining P and Q in the deleted plane is homotopic to $C_1 + n\Sigma$ where n is an integer (+,-,0). Hence we have the theorem:

<u>Theorem 7</u>. The line integral $\int_P^Q \sum_i v_i dx^i$ in the deleted plane is independent of the path iff:

$$(1)\quad \frac{\partial v_i}{\partial x^j} = \frac{\partial v_j}{\partial x^i} \quad \text{and} \quad (2)\quad \int_\Sigma \sum_i v_i dx^i = 0.$$

Thus we have a variation on Theorem 6 for this case:

<u>Theorem 6'</u>. Let $v_i(x)(i = 1,2)$ be C^1 functions defined in $R^2 - 0$. Then there exists a unique solution of the differential equation

$$\frac{\partial F}{\partial x^i} = v_i(x) \quad \text{such that} \quad F(x_o) = F_o$$

iff: (1) $\frac{\partial v_i}{\partial x^j} = \frac{\partial v_j}{\partial x^i}$ and (2) $\int_\Sigma \sum_i v_i dx^i = 0$ where Σ is a simple closed curve with the origin in its interior.

<u>Proof</u>. All that is needed to complete the proof is to show that if a solution exists, then $\int_\Sigma \sum_i v_i dx^i = 0$.

Under this hypothesis

$$\int_\Sigma \sum_i v_i dx^i = \int_\Sigma \sum_i \frac{\partial F}{\partial x^i} \frac{dx^i}{dt} = \int_\Sigma \frac{dF}{dt} = 0$$

<u>Example 1</u>. Let $v_1 = \frac{-y}{x^2+y^2}$; $v_2 = \frac{x}{x^2+y^2}$

The integrability conditions are satisfied, but

$\int_\Sigma \sum_i v_i dx^i = 2\pi$. Hence no solution F exists.

It might be thought that $F = \tan^{-1} \frac{y}{x}$ is a solution, for locally $\partial F/\partial x^i = v_i(x)$. The trouble is that $\tan^{-1}\frac{y}{x}$ is not defined as a single-valued function in the deleted plane.

Example 2. Let $v_1 = \frac{x}{x^2+y^2}$; $v_2 = \frac{y}{x^2+y^2}$

Then the integrability conditions are satisfied and $\int_\Sigma \sum_i v_i dx^i = 0$.

A solution is $\log\sqrt{(x^2+y^2)}$.

Remarks: (1) These ideas have an immediate extension to the case where D is the plane less any finite number of points.

(2) We shall see later (Chapter 7) that we have just proved an elementary form of the celebrated theorem of de Rham.

2.11 First Form of the Frobenius Theorem. This theorem is a generalization of Theorem 6 and is the most important theorem on differential equations in Differential Geometry.

Theorem 8. We are given the system of equations

$$(1) \quad \frac{\partial y^i}{\partial x^\alpha} = f^i_\alpha(y,x) \qquad i = 1\ldots n; \qquad \alpha = 1\ldots m$$

where y lies in an open set $V \subset R^n$ and x lies in an open, simply connected set $U \subset R^m$. We assume that $f^i_\alpha(y,x)$ are C^1 in $V \times U$. Let (y_0, x_0) be a point in $V \times U$ where $y_0 \varepsilon V$, and $x_0 \varepsilon U$. Then there exists an open, simply connected domain $\overline{U} \subset U$ containing x_0 such that in $\overline{U}$ there is a unique solution $y^i = g^i(x)$ with $y^i_o = g^i(x_0)$ if:

$$(2) \qquad \sum_j \frac{\partial f^i_\alpha}{\partial y^j} f^j_\beta + \frac{\partial f^i_\alpha}{\partial x^\beta} - \sum_i \frac{\partial f^i_\beta}{\partial y^j} f^j_\alpha - \frac{\partial f^i_\beta}{\partial x^\alpha} = 0 \text{ in } V \times U.$$

Note that if i ranges only over the value 1 and if $f_\alpha^1(y,x)$ are independent of y, this is the same as Theorem 6.

Proof. Let P be the point in U with coordinates x_0, and choose another point Q in U. (A restriction on the location of Q will appear shortly.) Construct a C^1 curve in U joining P and Q, with parametric equations $x = h(t)$ where $h(0)$ is P and $h(1)$ is Q. Along C consider the differential equation

$$\sum_\alpha \frac{\partial y^i}{\partial x^\alpha} \frac{dx^\alpha}{dt} = \sum_\alpha f_\alpha^i\,[y,x(t)]\,\frac{dx^\alpha}{dt}$$

By Theorem 1, this equation has a unique solution $y^i = g^i(t)$ with initial values y_0^i at $t = 0$ (or at P) over the whole range $0 \leq t \leq 1$, provided that Q is close enough to P. So we choose $\overline{U}\,(x_0)$ as a simply connected region of points having this property. We then define $y^i(Q) = g^i(1)$. In this way we have defined the functions $y^i(x)$ for $x \varepsilon \overline{U}$.

As in the proof of Theorem 6 we now require two further steps:

(a) To show that $y^i(Q)$ do not depend on the choice of curves joining P and Q; i.e. independence of the path.

(b) To show that $y^i(x)$ satisfy the given differential equation.

These proofs parallel those in Theorem 6.

(a) Independence of the path. Choose two curves joining P and Q, namely C_1 and C_2. Construct a C^{1+} homotopy: $x^\alpha = \phi^\alpha(\varepsilon,t)$ between C_1 and C_2. Then consider the differential equation

$$\text{(3)} \qquad \frac{dy^i}{dt} = \sum_\alpha f^i_\alpha\,[y,\phi(\varepsilon,t)]\,\frac{\partial\phi^\alpha}{\partial t}(\varepsilon,t)$$

For each value of ε we can integrate this equation and obtain a solution $y^i = g^i(\varepsilon,t)$. We wish to show that $g^i(\varepsilon,1)$ do not depend on ε. So we write

$$\text{(4)} \qquad \frac{\partial g^i(\varepsilon,t)}{\partial t} = \sum_\alpha f^i_\alpha\,[y,\phi(\varepsilon,t)]\,\frac{\partial\phi^\alpha}{\partial t}(\varepsilon,t)$$

and

$$\text{(5)} \qquad \frac{\partial g^i(\varepsilon,t)}{\partial \varepsilon} = \sum_\alpha f^i_\alpha\,[y,\phi(\varepsilon,t)]\,\frac{\partial\phi^\alpha}{\partial \varepsilon}(\varepsilon,t) + \sigma^i(\varepsilon,t)$$

where σ^i are so defined. From (4)

$$\frac{\partial^2 g^i(\varepsilon,t)}{\partial\varepsilon\partial t} = \sum_{\alpha,j} \frac{\partial f^i_\alpha}{\partial y^j}\left[\sum_\beta f^j_\beta \frac{\partial\phi^\beta}{\partial\varepsilon} + \sigma^j\right]\frac{\partial\phi^\alpha}{\partial t} + \sum_{\alpha,\beta}\frac{\partial f^i_\alpha}{\partial x^\beta}\frac{\partial\phi^\beta}{\partial\varepsilon}\frac{\partial\phi^\alpha}{\partial t}$$

$$+ \sum_\alpha f^i_\alpha \frac{\partial^2\phi^\alpha}{\partial\varepsilon\partial t}$$

and from (5)

$$\frac{\partial^2 g^i(\varepsilon,t)}{\partial t\partial\varepsilon} = \sum_{\alpha,j} \frac{\partial f^i_\alpha}{\partial y^j}\left[\sum_\beta f^j_\beta \frac{\partial\phi^\beta}{\partial t}\right]\frac{\partial\phi^\alpha}{\partial\varepsilon} + \sum_\alpha f^i_\alpha \frac{\partial^2\phi^\alpha}{\partial t\partial\varepsilon}$$

$$+ \frac{\partial\sigma^i}{\partial t} + \sum_{\alpha,\beta}\frac{\partial f^i_\alpha}{\partial x^\beta}\frac{\partial\phi^\beta}{\partial t}\frac{\partial\phi^\alpha}{\partial\varepsilon}$$

Subtracting we obtain:

$$0 = \sum_{j,\alpha,\beta}\left[\frac{\partial f^i_\alpha}{\partial y^j} f^j_\beta + \frac{\partial f^i_\alpha}{\partial x^\beta} - \frac{\partial f^i_\beta}{\partial y^j} f^j_\alpha - \frac{\partial f^i_\beta}{\partial x^\alpha}\right]\frac{\partial\phi^\alpha}{\partial\varepsilon}\frac{\partial\phi^\beta}{\partial t} + \frac{\partial\sigma^i}{\partial t}$$

$$- \sum_{j,\alpha}\frac{\partial f^i_\alpha}{\partial y^j}\sigma^j\frac{\partial\phi^\alpha}{\partial t}$$

By hypothesis the bracketed term is zero, and so

$$(6)\quad \frac{\partial\sigma^i}{\partial t} = \sum_{j,\alpha} \frac{\partial f^i_\alpha}{\partial y^j}\, \sigma^j\, \frac{\partial \bar{\phi}^\alpha}{\partial t}$$

Now examine (5) at $t = 0$. By our conditions it becomes

$$0 = 0 + \sigma^i(\varepsilon,0)$$

But for each $\varepsilon, \sigma^i(\varepsilon,t)$ satisfies the linear equation (6). Hence by Section 2.7 it follows that $\sigma^i(\varepsilon,t) \equiv 0$. Finally if we examine (5) at $t = 1$, we find

$$\frac{\partial g^i(\varepsilon,1)}{\partial \varepsilon} = 0$$

and so the independence of path is proved.

(b) The fact that $y^i(x)$ as so defined satisfy the differential equation is proved exactly as the corresponding fact for Theorem 6.

<u>References</u>

Birkhoff, G. and G.-C. Rota: <u>Ordinary Differential Equations</u> Ginn and Co., Boston, 1962. For the analytic case see pp. 119-121. Corollary 2.

Coddington, E. A. and N. Levinson: <u>Theory of Ordinary Differential Equations</u>, McGraw-Hill, New York, 1955. Chapter 1.

Ince, E. L.: <u>Ordinary Differential Equations</u>, Dover, New York, 1944. Chapter 3.

Murray, F. J. and K. S. Miller: <u>Existence Theorems for Ordinary Differential Equations</u>, New York University Press, New York, 1954. Chapter 1.

Thomas, T. Y.: <u>Systems of Total Differential Equations Defined Over Simply Connected Domains</u>, Annals of Mathematics, Vol. 35 (1934), pp. 730-734 for the Frobenius Theorem.

Exercises

1. Solve in the plane:

$$\frac{dx}{dt} = x + t - 1 \qquad \text{where} \quad x = 1 \quad \text{when} \quad t = 0.$$

by the method of successive approximations.

2. Show that $\frac{dx}{dt} = \sqrt{|x|}$ (in the plane) with initial conditions $(0,0)$ has the two solutions:

$$\text{(a)} \quad y = 0; \qquad \text{(b)} \quad y = \begin{cases} x^2/4 & x \geq 0 \\ -x^2/4 & x \leq 0 \end{cases}$$

Why does the Lipschitz Condition fail?

3. Prove Lemma 6, namely

$$|g_n(x_0 + s,t) - g_n(x_0,t)| \leq |s| \, [1 + (Kh) + \ldots + (Kh)^n]$$

on the basis of the hypothesis of Theorem 1.

4. In Theorem 8, where $\frac{\partial y^i}{\partial x^\alpha} = f^i_\alpha (y,x)$ compute (in terms of

$\left(f^i_\alpha, \frac{\partial f^i}{\partial y^j}, \frac{\partial f^i}{\partial x^\beta}\right)$ the expression: $\frac{\partial^2 y^i}{\partial x^\beta \partial x^\alpha} - \frac{\partial^2 y^i}{\partial x^\alpha \partial x^\beta}$

assuming the existence of a solution $y^i = g^i(x)$. The result should be the expression for the integrability conditions. Once this is realized, you should <u>never</u> in a specific case compute (2) by substituting in the formula. Instead compute

$\frac{\partial^2 y^i}{\partial x^\beta \partial x^\alpha} - \frac{\partial^2 y^i}{\partial x^\alpha \partial x^\beta}$ as above.

5. Consider the system

$$\frac{\partial x}{\partial u} = x - u - v \qquad \frac{\partial y}{\partial u} = -y + u + v$$

$$\frac{\partial x}{\partial v} = 1 \qquad \frac{\partial y}{\partial v} = 1$$

Show that the integrability conditions of the system are satisfied. If the initial conditions at $u = 0$, $v = 0$ are $x = 1$, $y = -1$, show that $x = u + v + 1$, $y = u + v - 1$ is a solution.

6. Consider the system:

$$\frac{\partial x}{\partial u} = x - u - v \qquad \frac{\partial y}{\partial u} = -y + u + v$$

$$\frac{\partial x}{\partial v} = x^2 - u^2 - v^2 - 2u - 2v - 2uv \qquad \frac{\partial y}{\partial v} = 1$$

Compute the integrability conditions. Note that these are not satisfied even though the system has the solution $x = u + v + 1$, $y = u + v - 1$. Hence equation (2) of Section (2.11) is not a necessary condition for the existence of a solution of (1). It is merely sufficient!

7. Find the integrability conditions of the system

$$\frac{\partial y^i}{\partial x^j} = -\sum_k \Gamma^i_{jk}(x) y^k \quad \text{where } \Gamma^i_{jk}(x) \text{ are } C^1 \text{ functions and}$$

i,j,k run from 1 to n. You have just computed the celebrated Riemann - Christoffel tensor!

8. Prove that the solution obtained in (2.11) is unique.

9. Find an analog of Theorem 6' for the system of Section (2.11).

Chapter 3

Theorems About Differentiable Functions

3.1 Introduction. In this chapter we shall discuss some of the important theorems about differentiable functions. The first of these, the Inverse Function Theorem, effectively says the following: Consider $f\colon R^n \to R^n$ near a point $\underline{a}$. Then if the affine approximation $A(x - a)$ to f near $\underline{a}$ is one-to-one (that is $\det df_a \neq 0$), then near $\underline{a}$ f is also one-to-one. The precise theorem, with all its fringes, is stated as Theorem 1.

Theorem 1. Given that $f\colon R^n \to R^n$ is of class C^p $(p > 0)$ in an open set S containing x_0 and that the Jacobian, J, of f at x_0 is not zero. Then there is an open set $U \subset S$ containing x_0 and an open set V containing $f(x_0)$ such that $f\colon U \to V$ has an inverse $h\colon V \to U$ of class C^p such that for all $y \in V$ $f \circ h$ = identity.

We give two proofs.

3.2 First Proof of the Inverse Function Theorem. To explain the ideas of this proof consider $f\colon R \to R$ of class C^1 defined in $S(x_0)$ and written $y = f(x)$. We assume $f'(x_0) \neq 0$ and so $f'(x) \neq 0$ and bounded in some $U(x_0)$ where $U \subset S$. Define $V = f(U)$.

Let x_1 and x_2 be any two points in U. Then by the ordinary mean value theorem

(1) $f(x_2) - f(x_1) = f'(\xi)(x_2 - x_1)$ for some ξ in U.

Hence $x_2 - x_1 \neq 0$ implies $f(x_2) - f(x_1) \neq 0$.

Thus $f: U \to V$ is one-to-one. This implies the existence of an inverse function $h: V \to U$ and that $f \circ h$ = identity. Let us write h in the form $x = h(y)$.

Then equation (1) can be rewritten as

$$(2) \quad y_2 - y_1 = f'(\xi)[h(y_2) - h(y_1)]$$

or

$$h(y_2) - h(y_1) = \frac{y_2 - y_1}{f'(\xi)}$$

Therefore h is continuous in V.

To prove that h is differentiable in V, consider

$$\lim_{y_2 \to y_1} \frac{h(y_2) - h(y_1)}{y_2 - y_1} = \lim_{x_2 \to x_1} \frac{1}{f'(\xi)} = \frac{1}{f'(x_1)}$$

Thus

$$\frac{dh}{dy} = \frac{1}{f'[g(y)]}$$

which is continuous in V, and h is of class C^1.

To get the rest of the theorem we observe that h is a solution of the differential equation:

$$\frac{dx}{dy} = \frac{1}{f'(x)} \qquad \text{with } h(y_0) = x_0$$

Hence if $f(x)$ is C^p, so is $g(y)$.

To extend this proof to $f: R^n \to R^n$ we use Theorem 12 of Chapter 1. Assume $S(x_0)$ to be convex. So for every two points x_1 and x_2 of S we have:

$$(3) \quad f^i(x_2) - f^i(x_1) = \sum_j g^i_j(x_2, x_1)(x^j_2 - x^j_1)$$

Consider the product space $S \times S$. At every point (x_1, x_2)

of $S \times S$ the $g^i_j\,(x_2,x_1)$ are defined and continuous, and $g^i_j(x_0,x_0) = (\partial f^i/\partial x^j)_{x_0}$. By hypothesis $\det\,(g^i_j(x_0,x_0) \neq 0$, and so there is a subset of $S \times S$, namely $U(x_0) \times U(x_0)$ within which $\det\,[g^i_j(x_2,x_1)] \neq 0$.

Hence for x_1 and x_2 in U, it follows from equation (3) that $f\colon U \to V$ is one-to-one. So the inverse function $h\colon V \to U$ exists and we have $x^i = h^i(y)$ for $y \in V$. Thus equation (3) becomes

$$\text{(4)} \qquad y^i_2 - y^i_1 = \sum_j [h^j(y_2) - h^j(y_1)]g^i_j\,(x_2,x_1)$$

or

$$\text{(5)} \qquad h^j(y_2) - h^j(y_1) = \sum_i (y^i_2 - y^i_1)(g^j_i)^{-1}\,(x_2,x_1)$$

Therefore

$$|h(y_2) - h(y_1)| \leq C|y_2 - y_1|$$

and h is continuous in V.

To prove that h is C^1 consider

$$\lim_{y_2 \to y_1} \frac{|h^j(y_2) - h^j(y_1) - \sum_i F^j_i(y^i_2 - y^i_1)|}{|y_2 - y_1|}$$

where F^j_i is the inverse of $\partial f^i/\partial x^j$ at x_1.

This is the same as

$$\lim_{y_2 \to y_1} \frac{|\sum_i [(g^j_i)^{-1}\,(x_2,x_1) - F^j_i]\,(y^i_2 - y^i_1)|}{|y_2 - y_1|} = 0$$

since $\lim_{x_2 \to x_1} (g_i^j)^{-1}(x_2, x_1) = F_i^j$

Thus

$$\frac{\partial h^j(y)}{\partial y^i} = F_i^j\,[h(y)]$$

which is continuous and so h is of class C^1. As before if f is of class C^p, so is h.

3.3 <u>Second Proof of the Inverse Function Theorem</u>. This proof relies on the Frobenius Theorem and assumes that f is class C^2 or higher.

We are given the equations $y^i = f^i(x)$ defined in S and that $\det(\partial f^i/\partial x^j) \neq 0$ at x_o. Hence $\det(\partial f^i/\partial x^j) \neq 0$ and is bounded in a neighborhood W of x_o where $W \subset S$, and $y_o = f(x_o)$.

To find the inverse function we try to solve:

(1) $$\frac{\partial x^j}{\partial y^i} = F_i^j(x) \qquad \text{where} \quad F_i^j(x) = (\partial f^i/\partial x^j)^{-1}$$

Equations (1) have a unique solution $x^i = h^i(y)$ satisfying $x_o = h(y_o)$ defined in a neighborhood $V \subset f(W)$ of $f(x_o)$ provided their integrability conditions are satisfied. To find these it is convenient to consider:

(2) $$\sum_j \frac{\partial x^j}{\partial y^i}\frac{\partial f^k}{\partial x^j} = \delta_i^k$$ which are equivalent to (1).

Then
$$\sum_j \frac{\partial^2 x^j}{\partial y^i \partial y^r}\frac{\partial f^k}{\partial x^j} + \sum_{j,s} \frac{\partial x^j}{\partial y^i}\frac{\partial^2 f^k}{\partial x^j \partial x^s}\frac{\partial x^s}{\partial y^r} = 0$$

So

$$\frac{\partial^2 x^j}{\partial y^i \partial y^r} = -\sum_k F^j_k \left[\sum_{j,s} \frac{\partial x^j}{\partial y^i} \frac{\partial^2 f^k}{\partial x^j \partial x^s} \frac{\partial x^s}{\partial y^r}\right]$$

Since the right side is symmetric in i and r, the integrability conditions are satisfied.

Thus we have functions $x^i = h^i(y)$ defined in V such that $x^i_o = h^i(y_o)$. To prove that in V we have $f \circ h$ = identity, we compute:

$$z^i = f^i[h(y)]$$

$$\frac{\partial z^i}{\partial y^j} = \sum_k \frac{\partial f^i}{\partial x^k} \frac{\partial h^k}{\partial y^j} = \delta^i_j$$

So

$$z^i = y^i + C^i$$

Since $h^i(y_o) = x^i_o$ and $f^i(x_o) = y^i_o$, it follows that $C^i = 0$ and that $f \circ h$ = identity. The set U of the theorem is therefore $U = h(V) \subset W$ so that $f(U) = V$.

3.4 Implicit Function Theorem. Closely related to the Inverse Function Theorem is the Implicit Function Theorem. The question here is that of whether the system of equations:

$$F_\alpha(x^1,\ldots,x^n,y^1,\ldots,y^m) \quad \alpha = 1\ldots m,\ i = 1\ldots n$$

can be solved for y^α in terms of x^i. To guess the result, we consider the corresponding linear system:

$$\sum_i a^\alpha_i x^i + \sum_\beta b^\alpha_\beta y^\beta = 0$$

From linear algebra we know that a unique solution for y^β exists if $\det(b^\alpha_\beta) \neq 0$. Hence we are led to generalize this to the theorem:

<u>Theorem 2</u>. Let $F_\alpha(x^1, \ldots, x^n, y^1, \ldots, y^m)$; $\alpha = 1\ldots m$, $i = 1\ldots n$ be m functions of class C^p $(p \geq 1)$ $F_\alpha: R^{n+m} \to R^m$ such that

(1) For $x_0 \in R^n$ and $y_0 \in R^m$: $F_\alpha(x_0, y_0) = 0$

(2) At (x_0, y_o) $\det(\partial F_\alpha / \partial y^\beta) \neq 0$.

$\partial x/\partial z = -\frac{\partial F/\partial z}{\partial F/\partial x}$

Then there exists an open set $U \subset R^n$ containing x_0 and an open set $V \subset R^m$ containing y_0 such that there exists a unique C^p function $f: U \to V$, $y = f(x)$, with $f(x_0) = y_0$ and $F_\alpha[x, f(x)] = 0$ for all x in U.

We give two proofs:

<u>First Proof</u>. This proof shows that the Implicit Function Theorem is an immediate consequence of the Inverse Function Theorem. Consider the system of equations:

$$(1)\quad \begin{cases} z^1 = x^1 \\ \vdots \\ z^n = x^n \\ z^{n+1} = F_1(x,y) \\ \vdots \\ z^{n+\alpha} = F_\alpha(x,y) \\ \vdots \\ z^{n+m} = F_m(x,y) \end{cases}$$

where
$$\begin{cases} z_o^i = x_o^i \\ z_o^{n+\alpha} = F_\alpha(x_o, y_o) = 0 \qquad \alpha = 1\ldots m \end{cases}$$

This system defines a function $\phi\colon R^{n+m} \to R^{n+m}$. Its Jacobian at (x_o, y_o) is

$$\det \left(\begin{array}{c|c} I & 0 \\ \hline \dfrac{\partial F_\alpha}{\partial x^i} & \dfrac{\partial F_\alpha}{\partial y^\beta} \end{array} \right)_{x_o, y_o} = \det \left(\frac{\partial F_\alpha}{\partial y^\beta} \right)_{x_o, y_o} \neq 0$$

Hence the Inverse Function Theorem applies and we have the inverse system:

$$(2) \quad \begin{cases} x^i = z^i \\ y^\alpha = h^\alpha (z^1 \ldots z^n, z^{n+1}, \ldots, z^{n+m}) \\ = h^\alpha (x^1, \ldots x^n, z^{n+1}, \ldots, z^{n+m}) \end{cases}$$

such that

$$(3) \quad F_\alpha [x, h(x,z)] = z^{n+\alpha} \quad \text{for all} \quad z^{n+\alpha}.$$

And in particular for $z^{n+\alpha} = z_o^{n+\alpha} = 0$, we have

$$F_\alpha [x, h(x,0)] = 0$$

So if we write $f(x) = h(x,0)$ we have the required function provided that $f(x_o) = y_o$. This follows from (2) since $y_o^\alpha = h^\alpha(x_o, 0) = f^\alpha(x_o)$.

Although the uniqueness of f can be inferred from this construction, a separate proof of uniqueness is more satisfying. Suppose that $f(x)$ and $g(x)$ both have the required properties:

$$F_\alpha[x,f(x)] = 0 \qquad\qquad F_\alpha[x,g(x)] = 0$$

$$f(x_o) = y_o \qquad\qquad g(x_o) = y_o$$

Then $y = f(x)$ and $y = g(x)$ are both solutions of the partial differential equations:

$$\frac{\partial F_\alpha}{\partial x^i} + \sum_\beta \frac{\partial F_\alpha}{\partial y^\beta} \frac{\partial y^\beta}{\partial x^i} = 0$$

or
$$\frac{\partial y^\beta}{\partial x^i} = - \sum_\alpha \frac{\partial F_\alpha}{\partial x^i} G^\beta_\alpha$$

where
$$G^\beta_\alpha = (\partial F_\alpha / \partial y^\beta)^{-1}$$

Moreover they have the same initial values. Hence they are equal.

Another form of the Implicit Function Theorem is as follows:

<u>Theorem 3</u>. Let $F_\alpha(z^1,\ldots,z^{n+m})$ be C^p $(p\geq 1)$ functions such that

(1) For some z_o, $F_\alpha(z_o) = 0$

(2) The matrix $\dfrac{\partial F_\alpha}{\partial z^A}$ is of rank m at z_o

where $A = 1\ldots n+m$.

Then there exist C^p functions $g: R^n \to R^{n+m}$ written $z^A = g^A(u^1,\ldots,u^n)$ defined in $U \subset R^n$ such that:

(1) $F_\alpha[g^A(u)] = 0$ for $u \in U$

(2) $g^A(u_o) = z^A_o$

(3) The matrix $\partial g^A/\partial u^i$ is of rank n in U.

Proof. Without loss of generality we assume that the minor:

$$\det \begin{pmatrix} \dfrac{\partial F_1}{\partial z^{n+1}} & \cdots & \dfrac{\partial F_1}{\partial z^{n+m}} \\ \vdots & & \vdots \\ \dfrac{\partial F_m}{\partial z^{n+1}} & \cdots & \dfrac{\partial F_m}{\partial z^{n+m}} \end{pmatrix}_{z_o} \neq 0$$

If we write $z^1 = x^1, \ldots z^n = x^n$, $z^{n+1} = y^1, \ldots z^{n+m} = y^m$, the conditions of Theorem 1 are satisfied. So we have:

$$y^\alpha = f^\alpha(x) \quad \text{or} \quad z^{n+\alpha} = f^\alpha(z^1, \ldots, z^n)$$

Thus our functions g can be taken to be:

$$\begin{cases} g^i(x) = x^i & i = 1 \ldots n \\ g^{n+\alpha}(x) = f^\alpha(x) & \alpha = 1 \ldots m \end{cases}$$

If we now write u^i for x^i, (1) and (2) of the theorem are proved. Part (3) is immediately evident, for the matrix $(\partial g^i/\partial x^j) = \delta^i_j$ or I.

Remark. In the sequel we shall think of u^i as parameters. The functions $g(u)$ are by no means unique. For we can introduce new parameters v by the equations $u^i = h^i(v)$ and then consider $g[h(v)] = \overline{g}(v)$ provided that at v_o, $\det(\partial u^i/\partial v^j) \neq 0$, where $u_0 = h(v_o)$.

3.5 Second Proof of the Implicit Function Theorem. There also exists a direct proof of this theorem that avoids the Inverse Function Theorem. Given $F_\alpha(x,y)$, a function $y = f(x)$ that satisfies $F[x,f(x)] = 0$ must also satisfy:

$$(1) \qquad \frac{\partial F_\alpha}{\partial x^i} + \sum_\beta \frac{\partial F_\alpha}{\partial y^\beta}\frac{\partial y^\beta}{\partial x^i} = 0$$

Hence we can find it, if it exists, by solving the differential equation:

$$(2) \qquad \frac{\partial y^\beta}{\partial x^i} = -\sum_\alpha \frac{\partial F_\alpha}{\partial x^i} G^\beta_\alpha$$

where the matrix $G^\beta_\alpha = (\partial F_\alpha/\partial y^\beta)^{-1}$, and hence exists in a neighborhood $U(x_0,y_0)$. So we need to check the integrability conditions of (2). Rather than doing so directly, we start from (1) and obtain:

$$\frac{\partial^2 F_\alpha}{\partial x^i \partial x^j} + \sum_\beta \frac{\partial^2 F_\alpha}{\partial x^i \partial y^\beta}\frac{\partial y^\beta}{\partial x^j} + \sum_\beta \frac{\partial^2 F_\alpha}{\partial y^\beta \partial x^j}\frac{\partial y^\beta}{\partial x^i} + \sum_{\beta,\gamma} \frac{\partial^2 F_\alpha}{\partial y^\beta \partial y^\gamma}\frac{\partial y^\beta}{\partial x^i}\frac{\partial y^\gamma}{\partial x^j}$$

$$+ \sum_\beta \frac{\partial F_\alpha}{\partial y^\beta}\frac{\partial^2 y^\beta}{\partial x^i \partial x^j} = 0$$

The first four terms taken as a whole are symmetric in i and j. Moreover $(\partial F_\alpha / \partial y^\beta)$ is a nonsingular matrix. Hence $\frac{\partial^2 y^\beta}{\partial x^i \partial x^j}$ is symmetric in i and j, and so the integrability conditions are satisfied.

We, therefore, choose $y^\beta = f^\beta(x)$ as solutions of (2) such that $y_0 = f(x_0)$. These are unique. By our choice of x_0 and y_0, $F[x_0,f(x_0)] = 0$. Moreover

$$\frac{DF_\alpha}{Dx^i} = \frac{\partial F_\alpha}{\partial x^i} + \sum_\beta \frac{\partial F_\alpha}{\partial y^\beta}\frac{\partial f^\beta}{\partial x^i} = 0$$

Hence for all x, $F[x,f(x)] = 0$.

3.6 <u>Local Submanifolds of R^n</u>. These intuitively are pieces of curves, surfaces, etc. that lie in R^n. Their study is important in itself and will motivate our treatment of differentiable manifolds in Chapter 6. Before treating the general case, let us study some familiar examples.

(1) <u>Curves in R^2</u>. There are three equivalent approaches:

(a) $F(x,y) = 0$ where F is C^1 and the matrix $\left(\frac{\partial F}{\partial x}\ \frac{\partial F}{\partial y}\right)$ is of rank 1 at (x_0,y_0) where $F(x_0,y_0) = 0$.

(b) $y = f(x)$ where f is C^1.

(c) $\begin{cases} x = p(t) \\ y = q(t) \end{cases}$ where the

where the matrix $\left(\frac{dp}{dt}\ \frac{dq}{dt}\right)$ is of rank 1.

These assume some restricted range of variables and thus are <u>local</u> definitions. The problem is to prove that they are <u>locally</u> equivalent.

If we start with (a) and apply Theorem (3), we obtain (c). If in (c) we apply Theorem 1 to $x = p(t)$, we obtain $t = p^{-1}(x) = r(x)$. [If $\frac{dp}{dt} = 0$, this does not work, so we use $y = q(t)$ and get $t = s(y)$.] Then put $t = r(x)$ in $y = q(t)$ to obtain $y = q[r(x)] = f(x)$, or (b). From (b) we get (a) by putting $F(x,y) = y - f(x)$. Thus, (a), (b) and (c) are equivalent.

(2) Curves in R^3. Then there are three equivalent approaches (locally):

(a) $$\begin{cases} F_1(x,y,z) = 0 \\ F_2(x,y,z) = 0 \end{cases}$$

where $F_1(x_0,y_0,z_0) = 0$, $F_2(x_0,y_0,z_0) = 0$ and the Jacobian matrix $\dfrac{\partial(F_1,F_2)}{\partial(x,y,z)}$ is of rank 2 at (x_0,y_0,z_0).

(b) $$\begin{cases} z = f_1(x) \\ y = f_2(x) \end{cases}$$

(c) $x = p(t)$, $y = q(t)$, $z = r(t)$ where the matrix $\left(\dfrac{dp}{dt}, \dfrac{dq}{dt}, \dfrac{dr}{dt}\right)$ is of rank 1.

(3) Surfaces in R^3. Again we have:

(a) $F(x,y,z) = 0$

where $F(x_0,y_0,z_0) = 0$ and the matrix $\left(\dfrac{\partial F}{\partial x}, \dfrac{\partial F}{\partial y}, \dfrac{\partial F}{\partial z}\right)$ is of rank 1 at (x_0,y_0,z_0).

(b) $z = f(x,y)$

(c) $x = p(u,v)$, $y = q(u,v)$, $z = r(u,v)$

where the Jacobian matrix $\dfrac{\partial(x,y,z)}{\partial(u,v)}$ is of rank 2 at (x_0,y_0,z_0) or at (u_0,v_0).

General Case.

(a) $F_\alpha(x) = 0$

where $\alpha = 1\ldots m < n$, $x \in R^n$, there is a solution $F_\alpha(x_0) = 0$, and the matrix $(\partial F_\alpha / \partial x^i)$ is of rank m at x_0.

$$\text{(b)} \quad \begin{cases} x^{n-m+1} = f^{n-m+1}(x^1, \ldots x^{n-m}) \\ \vdots \\ x^n \qquad = f^n(x^1, \ldots x^{n-m}) \end{cases}$$

(c) $x^i = g^i(u^1, \ldots, u^{n-m})$

where the matrix $(\partial g^i / \partial u^A)$ is of rank $n-m$ at u_0^A $(A = 1\ldots n-m)$.

From (a) we get (c) by the Implicit Function Theorem for $u \in U \subset R^{n-m}$. In (c) assume that $\det(\partial g^A/\partial u^B) \neq 0$. Then in $V \subset U$ we use the Inverse Function Theorem to obtain $u^A = h^A(x^1 \ldots x^{n-m})$. Then in $x^{n-m+1} = g^{n-m+1}(u^1, \ldots, u^{n-m})$ we substitute and obtain $x^{n-m+1} = g^{n-m+1}[h(x^1, \ldots, x^{n-m}] = f^{n-m+1}(x^1 \ldots x^{n-m})$ and similarly for the other equations in (b).

From (b) we obtain (a) by putting

$$\begin{cases} F_1(x) = x^{n-m+1} - f^{n-m+1}(x^1, \ldots, x^{n-m}) \\ \vdots \\ F_m(x) = x^n - f^n(x^1, \ldots, x^{n-m}) \end{cases}$$

We observe that one minor of the matrix $(\partial F/\partial x)$ is the identity, so this matrix is of rank m. Thus (a), (b) and (c) are equivalent.

The map ϕ defined by $x = g(u)$ in (c) is a C^p function $\phi: B \to R^n$ where B is an open ball in R^{n-m}. We now wish to investigate the properties of its image, W, which is a subset of R^n. From Theorem 12 of Chapter 1 there exist C^{p-1} functions $G^i_A(u_0 + h)$ such that

$$g^i(u_0 + h) - g^i(u_0) = \sum_i G^i_A(u_0 + h)\, h^A$$

where $G^i_A(u_0) = \left(\frac{\partial g^i}{\partial u^A}\right)_o$. Thus at u_0 the matrix G^i_A is of rank $n-m$, so there is an open ball $B_1 \subset B$ within which G^i_A remains of rank $n-m$. Therefore if $g(u_0 + h) = g(u_0)$ it follows that $h = 0$. That is $\phi | B_1$ is one-to-one.

Now choose a closed ball $\overline{B}_2 \subset B_1$ and an open ball $B_3 \subset \overline{B}_2$. The map $\tilde{\phi} = \phi | B_3$ is then a C^p, one-to-one, map $\tilde{\phi}: B_3 \to \tilde{W}$. We assert the result:

<u>Theorem 4</u>. $\tilde{\phi}$ is a homeomorphism. (For this to make sense we use the induced topology on W: namely the open sets of W are the intersections of W with the open sets of R^n.)

<u>Proof</u>. $\tilde{\phi}: B_3 \to W$ is one-to-one and continuous, and, in fact, extends to a one-to-one continuous map $\approx\!\!\phi: \overline{B}_2 \to \approx\!\!W$. We now quote the following theorem from topology:

<u>Lemma</u>. If $X \to Y$ is one-to-one and continuous, and if X is compact, then $Y \to X$ is continuous.

Therefore, since $\overline{B}_2$ is compact, $\approx\!\!\phi^{-1}$ is continuous and so $\tilde{\phi}^{-1}$ is continuous. Hence $\tilde{\phi}$ is a homeomorphism.

Later we shall wish to consider the concept of a manifold which is defined as follows:

Definition. A manifold is a topological space such that for every point, p, there is an open set $U(p)$ which is homeomorphic to an open ball in Euclidean space.

Theorem 4 thus shows that $\tilde{W}$ is a manifold.

Extension. Now consider $\emptyset$ defined by $x = g(u)$ in an open ball B where the matrix $(\partial g/\partial u)$ is everywhere of rank $n - m$. Then its image W is a manifold if $\emptyset$ is one-to-one. We say that W is an imbedded submanifold of R^n. It can happen, however, that $\emptyset$ is not one-to-one. For example W may be a self-interesting curve in the plane. Nevertheless its restrictions $\tilde{\emptyset}: B_3 \to \tilde{W}$ are one-to-one and $\tilde{W}$ is a manifold. In this case we say that W is an immersed submanifold of R^n.

3.7 Tangent and Normal Vectors for a Submanifold of R^n. Let M be an imbedded submanifold of R^n defined by the equivalent conditions:

(1) $x^i = g^i(u)$ where $\partial g^i/\partial u^A$ is of rank $n-m$.

(2) $F_\alpha(x) = 0$ where $\partial F_\alpha/\partial x^i$ is of rank m.

The usual restrictions on the domains of these functions are implied.

A curve on M is defined by a set of differentiable functions

$$u^A = f^A(t) \qquad A = 1 \ldots n - m$$

and thus has parametric equations in R^n

$$x^i = g^i[f(t)]$$

Its tangent vector at a point P has components (in R^n):

(1) $$\frac{dx^i}{dt} = \sum_A \frac{\partial g^i}{\partial u^A} \frac{du^A}{dt} \quad \text{at } P.$$

<u>Definition</u>. The tangent space to M at P is the set of vectors at P that are tangent to some differentiable curve on M passing through P.

<u>Theorem 5</u>. The tangent space to M at P is a linear space of dimension $n - m$.

<u>Proof</u>. For fixed A, $\partial g^i/\partial u^A$ is a tangent vector to the curve on M: $u^1 = c^1, \ldots, u^A = t, \ldots, u^{n-m} = c^{n-m}$. Moreover by hypothesis the set of vectors $\{\partial g^i/\partial u^A\}$ are independent since the matrix $(\partial g^i/\partial u^A)$ is of rank $n-m$. From equation (1) every tangent vector is linearly dependent on these $n - m$ independent vectors. Hence the tangent space has dimension $n - m$.

By definition, the normal space to M at P is the linear subspace of R^n at P which is the orthogonal complement of the tangent space at P.

<u>Theorem 6</u>. The normal space at P is spanned by the m independent vectors $\partial F_\alpha / \partial x^i$ at P where $\alpha = 1 \ldots m$.

<u>Proof</u>. We have $F_\alpha[g(u)] = 0$. So

$$\sum_i \frac{\partial F_\alpha}{\partial x^i} \frac{\partial g^i}{\partial u^A} = 0$$

Hence $\partial F_\alpha/\partial x^i$ for each $\alpha = 1 \ldots m$ are normal to M at P. The dimension of the normal space is $n - (n - m) = m$, and $\partial F_\alpha / \partial x^i$ are m independent vectors in this space. Hence they form a basis for the normal space.

Observe that $\partial g^i/\partial u^A$ and $\partial F_\alpha / \partial x^i$ together form a basis for R^n at P.

3.8 Extreme Values of $f: R^n \to R^1$.

Definition. Let $f: R^n \to R^1$ be defined in $U(x_o)$. Then f attains a relative maximum at x_o iff $f(x) < f(x_o)$ for all x near x_o. Similarly f attains a relative minimum at x_o iff $f(x) > f(x_o)$ for all x near x_o.

Theorem 7. Let f be differentiable in $U(x_o)$. Then if f attains a maximum or minimum at x_o, $(df)_{x_o} = 0$.

Proof. Let $\phi(t) = f(x_o + tv)$ for every t in some open interval of R^1 containing D. If x_o is a relative maximum for f, $\phi(t)$ has a relative maximum at $t = 0$. Hence from elementary calculus, $\phi'(0) = 0$. But $\phi'(0) = (df)_{x_o} v = 0$ for every v. Hence $(df)_{x_o} = 0$. Similarly for a minimum.

Definition. If $(df)_{x_o} = 0$, then x_o is a **critical** point of f.

Remark. Critical points include relative maxima and minima but may be of other kinds as well.

As in elementary calculus there is a second derivative test for relative maxima and minima:

Theorem 8. Let x_o be a critical point of $f: R^n \to R^1$ where f is C^2 at x_o. Then x_o is a relative minimum (maximum) if $\left(\frac{\partial^2 f}{\partial x^i \partial x^j}\right)_{x_o}$ is a positive (negative) definite matrix.

We recall that a matrix $A = (a_{ij})$ is positive (negative) definite if $\sum_{i,j} a_{ij} v^i v^j > 0$ (or < 0) for all nonzero vectors v.

Proof. (for a minimum). Since x_0 is critical there exist continuous functions $h_{ij}(x)$ such that

$$f(x) = f(x_0) + 1/2 \sum_{i,j} (x^i - x^i_o)(x^j - x^j_o)h_{ij}(x)$$

and $h_{ij}(x_0) = \left(\frac{\partial^2 f}{\partial x^i \partial x^j}\right)_{x_0}$.

If $\left(\frac{\partial^2 f}{\partial x^i \partial x^j}\right)_{x_0}$ is positive definite, then by continuity $h_{ij}(x)$ is positive definite in some $U(x_0)$. Hence

$$\sum_{i,j} (x^i - x^i_o)(x^j - x^j_o)\, h_{ij}(x) > 0 \quad \text{for} \quad x \in U(x_0).$$

Hence for all $x \in U$, $f(x) > f(x_0)$ and x_0 is a relative minimum. The proof for a maximum is similar.

3.9 Constrained Maxima and Minima. Consider a submanifold M of R^n defined by the m equations $F_\alpha(x) = 0$. Then M is of dimension $n - m$. Also consider a function $G: R^n \to R^1$, written $G(x)$, of class C^1. The restriction $G|M$ is then a function defined on M. We seek conditions that a point $P \in M$ be a critical point of $G|M$.

Suppose that x_0 is such a critical point. Then we can find parametric equations for M in $U(x_0) \subset M$:

$$x^i = f^i(u) \qquad \text{where} \quad x^i_o = f^i(u_0)$$

In terms of these $G|M$ has the expression $G[f(u)]$ which we continue to write $G(u)$. Since x_0 is a critical point of $G(u)$, by definition $\partial G/\partial u^A = 0$; $A = 1 \ldots n-m$. Thus

$$\frac{\partial G}{\partial u^A} = \sum_i \frac{\partial G}{\partial x^i} \frac{\partial f^i}{\partial u^A} = 0$$

Hence $\frac{\partial G}{\partial x^i}$ lies in the normal space to M at x_0. Since this normal space is spanned by $\partial F_\alpha / \partial x_i$, it follows that at x_0:

$$\frac{\partial G}{\partial x^i} = \sum_\alpha \lambda_\alpha \frac{\partial F_\alpha}{\partial x^i}$$

where λ_α are constants indicating the linear dependence.

We thus have the result:

Theorem 9. If P (with coordinates x_0) is a critical point of $G|M$, its coordinates satisfy:

$$(1) \quad \begin{cases} \dfrac{\partial G}{\partial x^i} = \sum_\alpha \lambda_\alpha \dfrac{\partial F_\alpha}{\partial x^i} \\ F_\alpha(x) = 0 \end{cases}$$

for some constants λ_α.

Conversely, if there is a set of λ_α such that (1) are satisfied at P, then P is a critical point of $G|M$. (Proof: reverse the above argument.)

Remark. The λ_α are called Lagrange multipliers.

The second derivative test to assure us that x_0 is a relative maximum or minimum is somewhat complicated. We treat the case of a minimum, and the case of a maximum is similar.

We know from Section 3.8 that x_0 is a minimum if the

matrix $\left(\frac{\partial^2 G}{\partial u^A \partial u^B}\right)_{x_0}$ is positive definite. Now

$$\frac{\partial^2 G}{\partial u^A \partial u^B} = \sum_{i,j} \frac{\partial^2 F_\alpha}{\partial x^i \partial x^j} \frac{\partial f^i}{\partial u^A} \frac{\partial f^j}{\partial u^B} + \sum_i \frac{\partial G}{\partial x^i} \frac{\partial^2 f^i}{\partial u^A \partial u^B}$$

Moreover:

$$\frac{\partial^2 F_\alpha}{\partial u^A \partial u^B} = \sum_{i,j} \frac{\partial^2 F_\alpha}{\partial x^i \partial x^j} \frac{\partial f^i}{\partial u^A} \frac{\partial f^j}{\partial u^B} + \sum_i \frac{\partial F_\alpha}{\partial x^i} \frac{\partial^2 x^i}{\partial u^A \partial u^B} \equiv 0$$

Hence, using equations (1), by subtraction;

$$(2) \quad \left(\frac{\partial^2 G_\alpha}{\partial u^A \partial u^B}\right)_{x_0} = \sum_{i,j}\left(\frac{\partial^2 G}{\partial x^i \partial x^j} - \sum_\alpha \lambda_\alpha \frac{\partial^2 F_\alpha}{\partial x^i \partial x^j}\right)_{x_0} \frac{\partial f^i}{\partial u^A} \frac{\partial f^j}{\partial u^B}$$

Let $T_{ij} = \left(\frac{\partial^2 G}{\partial x^i \partial x^j} - \sum_\alpha \lambda_\alpha \frac{\partial^2 F_\alpha}{\partial x^i \partial x^j}\right)_{x_0}$.

Therefore P is a minimum if

$$\sum_{i,j} T_{ij} \frac{\partial x^i}{\partial u^A} \frac{\partial x^j}{\partial u^B}$$

is positive definite. From this we derive Theorem 10.

<u>Theorem 10</u>. A critical point, P, of $G|M$ is a relative minimum if the matrix T_{ij} is positive definite.

This works well for many applications, but in other cases a more sophisticated test is required to settle the question. Since $\partial f^i/\partial u^A$ are tangent vectors to M, we derive from equation (2) that P is a minimum if

$$\sum_{i,j} T_{ij} V^i V^j > 0 \text{ for all vectors } V^i \text{ such that}$$

$$\sum_{i,j} \delta_{ij} V^i V^j = 1 \qquad \text{and} \qquad \sum_i \frac{\partial F_\alpha}{\partial x^i} V^i = 0$$

These conditions state that V is a unit tangent vector to M.

So we consider the function $\sum_{i,j} T_{ij} V^i V^j$ subject to the constraints just stated on V. That is, $\sum_{i,j} T_{ij} V^i V^j$ is defined on the unit sphere in the R^{n-m} tangent to M at x_o. To show that it is positive at all points on this sphere, we prove that its minimum value on this sphere is positive. Now this minimum is attained at a critical point; so if $\sum_{i,j} T_{ij} V^i V^j$ is positive at all its critical points, we are certain that it is positive on the entire sphere. Hence we look for the critical points.

By the method of Lagrange we are led to the equation:

$$(3) \qquad 2 \sum_j T_{ij} V^j = 2\sigma \sum_j \delta_{ij} V^j + \sum_\alpha \rho_\alpha \frac{\partial F_\alpha}{\partial x^i}$$

where now σ and ρ_α are the Lagrange multipliers.

From (3) we see that

$$2 \sum_{i,j} T_{ij} V^i V^j = 2\sigma \sum_{i,j} V^i V^j + 0$$

So

$$\sigma = \sum_{i,j} T_{ij} V^i V^j$$

Now consider (3) together with $\sum_i \frac{\partial F_\alpha}{\partial x^i} V^i = 0$ as $n+m$ equations in R^{n+m} with unknowns $V^1, \ldots, V^n, \rho_1, \ldots, \rho_m$. We know that there is a nontrivial solution of these equations (for a minimum surely exists since the sphere is compact). Hence the following determinant must be zero:

$$\left|\begin{matrix} T_{11}-\sigma & T_{12} & \cdots & T_{1n} & \frac{\partial F_1}{\partial x^1} & \cdots & \frac{\partial F_m}{\partial x^1} \\ \vdots & T_{22}-\sigma & & \vdots & \vdots & & \vdots \\ T_{n1} & --- & & T_{nn}-\sigma & \frac{\partial F_1}{\partial x^n} & \cdots & \frac{\partial F_m}{\partial x^n} \\ \frac{\partial F_1}{\partial x^1} & --- & & \frac{\partial F_1}{\partial x^n} & 0 & \cdots & 0 \\ \vdots & & & \vdots & \vdots & & \vdots \\ \frac{\partial F_m}{\partial x^1} & --- & & \frac{\partial F_m}{\partial x^n} & 0 & \cdots & 0 \end{matrix}\right|_{x_0} = 0$$

This is a polynomial of degree $n - m$ in σ. If its roots, σ, are all positive, $\sum_{i,j} T_{ij}v^iv^j$ is positive at every critical point. Thus P (with coordinates x_0) is a minimum. If the roots are all negative, P is a maximum.

3.10 Diagonalization of a Symmetric Matrix. This is a standard problem in Linear Algebra, but the proofs there are either omitted or have no conceptual content. The problem is introduced here as an interesting application of the method of Lagrange.

Theorem 11. Given a real symmetric $n \times n$ matrix A, there exists an orthogonal matrix M such that $M^T A M = D$ where D is a diagonal matrix of the form

$$D = \begin{pmatrix} \lambda_1 & & 0 \\ & \ddots & \\ 0 & & \lambda_n \end{pmatrix}$$ where λ_i are real numbers.

Proof. Consider $f(X) = X^T AX$ where $X^T IX = 1$ is a function on the unit sphere S^{n-1} in R^n. Then f must have a maximum value on S^{n-1} since S^{n-1} is compact. The point X_1 at which this is attained must satisfy (by Lagrange):

$$AX_1 - \lambda_1 IX_1 = 0 \qquad \text{or} \quad (A - \lambda_1 I)X_1 = 0$$

Hence λ_1 is a characteristic root of A. Moreover, $\lambda_1 = X_1{}^T AX_1$, and so λ_1 is the maximum value of $f(X)$ on S^{n-1}.

Next consider $f(X)$ restricted to the unit sphere S^{n-2} in R^{n-1} orthogonal to X_1. The values of X are now constrained by $X^T IX = 1$ and $X^T IX_1 = 0$. Again $f(X)$ has a maximum value on S^{n-2} which is attained at a point X_2. Then X_2 must satisfy:

$$AX_2 - \lambda_2 IX_2 - \sigma_1 IX_1 = 0$$

where λ_2 and σ_1 are Lagrange multipliers. Then

$$X_2^T AX_2 - \lambda_2 X_2^T IX_2 = 0, \quad \text{or} \quad \lambda_2 = X_2^T AX_2.$$

So λ_2 is the maximum value of $f(X)$ on S^{n-2}, and hence $\lambda_1 \geq \lambda_2$. Moreover:

$$X_1^T AX_2 - \lambda_2 X_1^T IX_2 - \sigma_1 = 0 \quad \text{or} \quad \sigma_1 = X_1^T AX_2$$

We know, however, that $X_1^T A = \lambda_1 X_1^T I$ (since A is symmetric). So

$$\sigma_1 = X_1^T AX_2 = \lambda_1 X_1^T IX_2 = 0; \quad \text{or} \quad \sigma_1 = 0$$

Thus X_2 satisfies $(A - \lambda_2 I)X_2 = 0$ and λ_2 is a characteristic root of A.

More generally, suppose that by this process we have found $k-1(\leq n-2)$ unit mutually orthogonal vectors $X_1 \ldots X_{k-1}$ and corresponding characteristic roots of A:

$$\lambda_1 \geq \lambda_2 \geq \cdots \geq \lambda_{k-1}$$

We now consider $f = X^T AX$ restricted to the unit sphere S^{n-k} lying in R^{n-k+1} orthogonal to all of $X_1 \ldots X_{k-1}$. Again f has a maximum value on S^{n-k} attained at X_k. By Lagrange X_k satisfies:

$$AX_k - \lambda_k IX_k - \sum_{\alpha=1}^{k-1} \sigma_\alpha IX_\alpha = 0$$

where $\lambda_k, \sigma_1, \ldots, \sigma_{k-1}$ are Lagrange multipliers. Then

$$X_k^T AX_k = \lambda_k$$

So λ_k is the maximum of $f(X)$ on S^{n-k} and

$$\lambda_1 \geq \lambda_2 \geq \cdots \geq \lambda_{k-1} \geq \lambda_k$$

Moreover:

$$X_\alpha^T A X_k - \lambda_k X_\alpha IX_k = \sigma_\alpha; \quad \text{or} \quad \sigma_\alpha = X_\alpha^T AX_k$$

We know, however, that $X_\alpha^T A = \lambda_\alpha X_\alpha^T I$; so

$$\sigma_\alpha = X_\alpha^T AX_k = \lambda_\alpha X_\alpha^T IX_k = 0.$$

Thus X_k satisfies $(A - \lambda_k I)X_k = 0$ and λ_k is a characteristic root of A.

This process stops when $k = n - 1$, for at the next step we must consider f restricted to $S^{n-(n-1)-1} = S^o$; and this is not differentiable. Nevertheless, there is a maximum at X_n which is a unit vector orthogonal to $X_1 \ldots X_{n-1}$. Define $\lambda_n = X_n^T AX_n$, the maximum value of f on S^o. Then $\lambda_1 \geq \lambda_2 \geq \ldots \geq \lambda_n$.

Now AX_n is a vector in R^n and hence

$$AX_n = \sum_{i=1}^{n} c_i X_i$$

where

$$c_\beta = X_\beta^T AX_n = \lambda_\beta X_\beta^T IX_n = 0; \quad \beta = 1 \ldots n - 1$$

and

$$c_n = \lambda_n.$$

Hence $(A - \lambda_n I)X_n = 0$ and λ_n is a characteristic root of A.

Now we have an orthonormal frame $X_1 \ldots X_n$ and characteristic roots $\lambda_1 \ldots \lambda_n$ such that $AX_i = \lambda_i X_i$ and

$$X_i^T AX_j = \begin{cases} \lambda_i & i = j \\ 0 & i \neq j \end{cases}$$

Choose M as the matrix whose column vectors are X_i.

$$\text{Then } M^T AM = (X_i^T AX_j) = \begin{pmatrix} \lambda_1 & & 0 \\ & \ddots & \\ 0 & & \lambda_n \end{pmatrix} = D.$$

3.11 Frobenius Theorem (Second Form)

Preliminary. Suppose that in an open set U of R^n we are given a C^1 vector field X. This means that the components of

X, namely X^i, are C^1 functions $R^n \to R^1$ defined in $U \subset R^n$. Problem: Can we find a family of curves in R^n such that one curve of the family passes through each point of U and such that in U, X is tangent to the curves of the family? In other words, can we solve:

$$\frac{dx^i}{dt} = X^i(x)$$

with given initial conditions? From the existence theorem, the answer is clearly "yes".

<u>Frobenius</u>. Now suppose that we have $r(\geq 2)$ independent vector fields X_A $(A = 1...r)$ in $U \subset R^n$. Does there exist a family of submanifolds of dimension $r, M^r \subset R^n$ such that one manifold of the family passes through each point of U and such that X_A are tangent to the manifolds of the family?

The problem can be restated: Do there exist $n - r$ functions $R^n \to R$ namely $F_\alpha(\alpha = 1...n - r)$ with $\partial F_\alpha / \partial x^i$ of rank $n - r$ such that

$$\sum_i X_A^i \frac{\partial F_\alpha}{\partial x^i} = 0$$

where X_A^i are the components of X_A?

For $F_\alpha(x)$ = constant defines an M^r to which $\partial F_\alpha / \partial x^i$ are normal. Thus X_α^i are tangent.

In order to state the Frobenius Theorem, we must define the Lie Bracket $[X_A, X_B]$ of two vector fields. This is a somewhat awkward definition which will be simplified in Chapter 6.

<u>Definition</u>. The Lie Bracket $[X_A, X_B]$ of two vector fields in R^n is the vector field whose component functions are:

$$[X_A, X_B]^i = \sum_j \left(X_A^j \frac{\partial X_B^i}{\partial x^j} - X_B^j \frac{\partial X_A^i}{\partial x^j} \right)$$

Theorem 12. (Frobenius) Given $r(\geq 2)$ independent $C^p(p \geq 2)$ vector fields in an open set U of R^n. Then through each point of U there exists a submanifold M^r to which X_A are tangent iff $[X_A, X_B] = \sum_C C_{AB}^C X_C$ where C_{AB}^C are functions $R^n \to R^1$ defined in U.

Proof. Necessity. If the required $F_\alpha(x)$ exist, then

$$0 = \sum_j X_A^j \frac{\partial}{\partial x^j} \left(\sum_i X_B^i \frac{\partial F_\alpha}{\partial x^i} \right) = \sum_{i,j} X_A^j \frac{\partial X_B^i}{\partial x^j} \frac{\partial F_\alpha}{\partial x^i} + \sum_{i,j} X_A^j X_B^i \frac{\partial^2 F_\alpha}{\partial x^i \partial x^j}$$

Interchanging A and B and subtracting we get

$$\sum_{i,j} \left(X_A^j \frac{\partial X_B^i}{\partial x^j} - X_B^j \frac{\partial X_A^i}{\partial x^j} \right) \frac{\partial F_\alpha}{\partial x^i} = 0$$

Hence the vector $[X_A, X_B]$ on the left is a tangent vector to M^r and so is expressible as a linear combination of the basis for tangent vectors:

$$[X_A, X_B] = \sum_C C_{AB}^C X_C$$

This is true on every M^r of the family and hence in U.

Sufficiency. We begin with a lemma:

Lemma 1. If $[X_A, X_B] = \sum_C C_{AB}^C X_C$ and $Y_B = \sum_A a_B^A X_A$ where the matrix (a_B^A) is nonsingular, then

$$[Y_A, Y_B] = \sum_C \overline{C}^C_{AB} Y_C \quad \text{for some } \overline{C}^C_{AB}.$$

This lemma is obvious if we have proved Frobenius; but here we must prove it the long way by direct computation:

Proof:

$$[Y_A, Y_B]^i = \sum_{C,j} a^C_A X^j_C \frac{\partial\left(\sum_D a^D_B X^i_D\right)}{\partial x^j} - \sum_{C,j} a^C_B X^j_C \frac{\partial\left(\sum_D a^D_A X^i_D\right)}{\partial x^j}$$

$$= \sum_{j,C,D} a^C_A a^D_B \left[X^j_C \frac{\partial X^i_D}{\partial x^j} - X^j_D \frac{\partial X^i_C}{\partial x^j}\right] + \sum_{j,C,D} \left[a^C_A \frac{\partial a^D_B}{\partial x^j} - a^C_B \frac{\partial a^D_A}{\partial x^j}\right] X^j_C X^i_D$$

$$= \sum_{C,D,E} \left[a^C_A a^D_B C^E_{CD} X^i_E\right] + \sum_{j,C,E} X^i_E \left[a^C_A \frac{\partial a^E_B}{\partial x^j} - a^C_B \frac{\partial a^E_A}{\partial x^j}\right] X^j_C$$

$$= \sum_{C,D,E,j} X^i_E \left[a^C_A a^D_B C^E_{CD} + \left\{a^C_A \frac{\partial a^E_B}{\partial x^j} - a^C_B \frac{\partial a^E_A}{\partial x^j}\right\} X^j_C\right]$$

$= \sum_E X^i_E D^E_{AB}$ where D^E_{AB} are the bracket expressions above.

Hence $[Y_A, Y_B]$ is linearly dependent on the set of vectors $\{X_A\}$ and thus (since a^B_C is nonsingular) on the set of vectors $\{Y_A\}$. That is:

$$[Y_A, Y_D] = \sum_C \overline{C}^C_{AD} Y_C \quad \text{for some } \overline{C}^C_{AB}.$$

Lemma 2. Given the set of vectors X_A, there exists a nonsingular transformation $Y_B = \sum_A a^A_B X_A$ such that $[Y_A, Y_B] = 0$.

Proof. Since the vectors X^i_A are independent, there exists a nonzero minor of the matrix (X^i_A) which can be assumed to be

(X^B_A). Then define $(a^A_B) = (X^A_B)^{-1}$. Hence $(Y^A_B) = (\delta^A_B)$.

From the first line of the proof of Lemma 1 we see that

$$[Y_A, Y_B]^C = 0$$

But from Lemma 1, we know that

$$0 = [Y_A, Y_B]^C = \sum_E \overline{C}^E_{AB} Y^C_E = \overline{C}^C_{AB}. \quad \text{Thus} \quad [Y_A, Y_B] = 0.$$

To prove the theorem, we now choose Y^i_A as in Lemma 2 and consider the equations:

$$\frac{\partial x^i}{\partial u^A} = Y^i_A(x)$$

whose integrability conditions are $[Y_A, Y_B] = 0$. Since these are satisfied, there exists a submanifold, M^r, of R^n passing through a chosen point of U to which Y^i_A (and hence X^i_A) are tangent.

3.12 Third Form of the Frobenius Theorem (Preliminary Version).
A related problem is the following: Given a vector field in R^n, say X; is there a family of $n-1$ dimensional submanifolds normal to X at each point? In other words, is there a function $F(x)$ and a factor of proportionality $\lambda(x)$ such that

$$\frac{\partial F}{\partial x^i} = \lambda X_i$$

This problem is then a generalization of the problem studied in Section 2.9.

Another formulation of this problem is in the language of Total Differential Equations. The expression

$$\sum_i X^i(x)dx^i = 0 \qquad i = 1 \dots n$$

is called a total differential equation. In R^3 it is usually written

$$Pdx + Qdy + Rdz = 0.$$

A solution of a total differential equation is a submanifold of dimension $n-1$ which is normal to X at each point. The question is whether such a solution exists.

We do not yet have the machinery to answer this question in general, but we can answer it here in the important special case of R^3. The relevant theorem is:

<u>Theorem 13</u>. If X is a vector field defined in an open set U of R^3, there is a two-dimensional submanifold of R^3 passing through each point of U and normal to X, iff in U:

$$X \cdot \operatorname{curl} X = 0$$

<u>Proof</u>. <u>Necessity</u>. Suppose such a submanifold M exists through a point $x_0 \in U$. Then choose two independent vector fields in U, A and B, such that $A \wedge B = X$ (here $\wedge$ indicates the vector product) and A and B are tangent to M.

Then by the second form of Frobenius:

(1) $$\sum_i A^i \frac{\partial B^j}{\partial x^i} - B^i \frac{\partial A^j}{\partial x^i} = c_1 A^j + c_2 B^j$$

Since $A \cdot X = 0$, it follows that

(2) $$\sum_{i,j} A^i \frac{\partial B^j}{\partial x^i} X_j - B^i \frac{\partial A^j}{\partial x^i} X_j = 0$$

Now $\displaystyle\sum_j \frac{\partial A^j}{\partial x^i} X_j = - \sum_j A^j \frac{\partial X_j}{\partial x^i}$ and similarly for B.

and that $(x_0, y_0, z_0, u_0, v_0, w_0)$ satisfies the system. Under what conditions can we solve and obtain

$$u = f(x,y,z)$$
$$v = g(x,y,z)$$
$$w = h(x,y,z)$$

near (x_0, y_0, z_0)? Solve explicitly for u^2, v^2 and w^2 and explain why your conditions found above are necessary.

5. Let $G = x - y + 2z$ in R^3

$$F = x^2 + y^2 + 2z^2$$

Then $F = 2$ defines an M^2 in R^3 (an ellipsoid). Find the critical points of $G|M$. Determine whether maximum or minimum.

Ans. $\pm \frac{\sqrt{2}}{2}(1,-1,1)$.

6. Consider the problem: Find the critical points of $G(x,y) = y$ subject to the constraint $F(x,y) = (y - x^2)(2y - x^2) = 0$. The Lagrange method fails, but $y = 0$ is an obvious minimum point for $G(x,y)$. What goes wrong?

7. Find the critical point of $G(x,y,z) = xyz$ subject to the constraint $F(x,y,z) = \frac{1}{x} + \frac{1}{y} + \frac{1}{z} = c$ where $x > 0$, $y > 0$, $z > 0$. Is this a maximum, minimum or neither?

8. Prove that the shortest distance from a point to a line in R^3 is the perpendicular distance.

$$\sum_i X^i(x)dx^i = 0 \qquad i = 1\ldots n$$

is called a total differential equation. In R^3 it is usually written

$$Pdx + Qdy + Rdz = 0.$$

A solution of a total differential equation is a submanifold of dimension $n-1$ which is normal to X at each point. The question is whether such a solution exists.

We do not yet have the machinery to answer this question in general, but we can answer it here in the important special case of R^3. The relevant theorem is:

<u>Theorem 13</u>. If X is a vector field defined in an open set U of R^3, there is a two-dimensional submanifold of R^3 passing through each point of U and normal to X, iff in U:

$$X \cdot \text{curl } X = 0$$

<u>Proof</u>. <u>Necessity</u>. Suppose such a submanifold M exists through a point $x_0 \in U$. Then choose two independent vector fields in U, A and B, such that $A \wedge B = X$ (here $\wedge$ indicates the vector product) and A and B are tangent to M.

Then by the second form of Frobenius:

(1) $$\sum_i A^i \frac{\partial B^j}{\partial x^i} - B^i \frac{\partial A^j}{\partial x^i} = c_1 A^j + c_2 B^j$$

Since $A \cdot X = 0$, it follows that

(2) $$\sum_{i,j} A^i \frac{\partial B^j}{\partial x^i} X_j - B^i \frac{\partial A^j}{\partial x^i} X_j = 0$$

Now $\sum_j \frac{\partial A^j}{\partial x^i} X_j = - \sum_j A^j \frac{\partial X_j}{\partial x^i}$ and similarly for B.

Thus

$$\sum_{i,j} A^i B^j \frac{\partial X_j}{\partial x^i} - B^i A^j \frac{\partial X_j}{\partial x^i} = 0$$

or

$$\sum_{i,j} (A^i B^j - B^i A^j) \left(\frac{\partial X_j}{\partial x^i} - \frac{\partial X_i}{\partial x^j} \right) = 0$$

In vector notation this becomes:

(3) $\quad (A \wedge B) \cdot \text{curl } X = 0, \quad \text{or} \quad X \cdot \text{curl } X = 0$

<u>Sufficiency</u>. We assume $X \cdot \text{curl } X = 0$ and choose A and B so that $A \wedge B = X$. Then we reverse the argument to obtain (2). But (2) says that

$$\sum_i A^i \frac{\partial B^j}{\partial x^i} - B^i \frac{\partial A^j}{\partial x^i}$$

is orthogonal to X and hence is a linear combination of A and B. This gives (1). Now apply the second form of Frobenius to find the submanifold to which A and B are tangent, or to which X is normal.

Exercises

1. For the following functions: Compute the Jacobian. Where is the function one-to-one? Find its inverse where it exists.

   (a) $x = \frac{2u}{u^2+v^2}, \quad y = -\frac{2v}{u^2+v^2}$

   (b) $x = u^2 - v^2, \quad y = 2uv$

   (c) $x = \frac{u}{1-u-v}, \quad y = \frac{v}{1-u-v}$

2. Let $f: R^2 \to R^2$ be given by:

   $$x^1 = (\exp s)(\cos t)$$

   $$x^2 = (\exp s)(\sin t)$$

   Show that f satisfies the hypotheses of the inverse function theorem at every point of R^n, but that f is not one-to-one.

   If f is restricted to $\{(s,t) \mid 0 < t < 2\pi\}$ show that it is one-to-one and find its inverse.

3. Let $F = x^2 + y^2 + z^2 - 2xz - 4$ in R^3.
   Prove that $F = 0$ defines an M^2 in the neighborhood of any point (x_0, y_0, z_0) for which $F(x_0, y_0, z_0) = 0$.

4. Given the system

$$\begin{cases} u^2 + v^2 + w^2 - x^2 = 0 \\ u^2 + v^2 - y^2 = 0 \\ u^2 + w^2 - z^2 = 0 \end{cases}$$

and that $(x_o, y_o, z_o, u_o, v_o, w_o)$ satisfies the system. Under what conditions can we solve and obtain

$$u = f(x,y,z)$$
$$v = g(x,y,z)$$
$$w = h(x,y,z)$$

near (x_o, y_o, z_o)? Solve explicitly for u^2, v^2 and w^2 and explain why your conditions found above are necessary.

5. Let $G = x - y + 2z$ in R^3

$$F = x^2 + y^2 + 2z^2$$

Then $F = 2$ defines an M^2 in R^3 (an ellipsoid). Find the critical points of $G|M$. Determine whether maximum or minimum.

Ans. $\pm \frac{\sqrt{2}}{2}(1,-1,1)$.

6. Consider the problem: Find the critical points of $G(x,y) = y$ subject to the constraint $F(x,y) = (y - x^2)(2y - x^2) = 0$. The Lagrange method fails, but $y = 0$ is an obvious minimum point for $G(x,y)$. What goes wrong?

7. Find the critical point of $G(x,y,z) = xyz$ subject to the constraint $F(x,y,z) = \frac{1}{x} + \frac{1}{y} + \frac{1}{z} = c$ where $x > 0$, $y > 0$, $z > 0$. Is this a maximum, minimum or neither?

8. Prove that the shortest distance from a point to a line in R^3 is the perpendicular distance.

9. Find the minimum distance from the origin to the surface $(x - y)^2 - z^2 = 1$.

10. Consider the $M^2 \subset R^3$ whose equation is the central quadric:

$$F(x) = \sum_{i,j=1}^{3} a_{ij}\, x_i^i x^j - 1 = 0$$

Find those points on M^2 at which the square of the distance to the origin: $d^2 = \sum_{i,j=1}^{3} \delta_{ij} x^i x^j$ has a critical value.

11. Prove the fact used in the proof of Theorem 10: If T_{ij} is positive definite, then

$$\sum_{i,j} T_{ij} \frac{\partial x^i}{\partial u^A} \frac{\partial x^j}{\partial u^B} \quad \text{is positive definite.}$$

12. Prove that the following vector fields in R^3 satisfy the hypotheses of the Frobenius Theorem and find the equation of the corresponding family of submanifolds $M^2 \subset R^3$. Verify that X_1 and X_2 are tangent to M^2.

$$X_1 = (1,0,0); \quad X_2 = (x,y,z) \quad \text{where} \quad y > 0 \text{ and } z > 0.$$

13. Same as exercise 12 with

$$X_1 = (x,y,0); \quad X_2 = (0,y,z) \quad \text{where} \quad x > 0,\ y > 0,\ z > 0.$$

14. Is there a family of surfaces in R^2 orthogonal to:

(a) $(yz,\ 2zx,\ -3xy)$; (b) $(z, x + y,\ 1)$?

15. Solve the total differential equation:

$$y\, z \log z\, dx - zx \log z\, dy + xy\, dz = 0$$

16. A map f: $R^n \to R^n$ is "proper" if $f^{-1}(K)$ is compact whenever K is compact. Prove: A C^1 map f: $R^n \to R^n$ is a diffeomorphism iff it is **proper** and its Jacobian is nowhere zero. (See W. B. Gordon: "On the Diffeomorphisms of Euclidean Space", American Mathematical Monthly, Vol. 79 (1972) pp. 755-759; and Vol. 80 (1973) pp. 674-675.

Chapter 4

The Geometry of Submanifolds

4.1 Introduction. In this chapter we turn to the study of the differential geometry of submanifolds of R^n. This will serve as an immediate application of the previous chapters and as motivation for the study of abstract differentiable manifolds in Chapters 6 and 7. Although the dimension of a submanifold can be any integer between one and $n - 1$, there are complications in the general case that we wish to avoid. The essential ideas appear in submanifolds of dimension one (curves) and those of dimension $n - 1$ (hypersurfaces); and so we proceed to consider these in turn.

4.2 Definition of a Curve in R^n. Consider an interval $I: a \leq t \leq b$ in R^1 and a function $f: I \to R^n$. If a point in R^n is given by the coordinates $x^1, x^2, \ldots, x^n$, f can be written as a set of parametric equations:

$$x^1 = f^1(t)$$

$$x^2 = f^2(t)$$

$$\vdots$$

$$x^n = f^n(t)$$

or more simply as

$$x^i = f^i(t) \qquad i = 1 \ldots n$$

or $\quad x = f(t) \quad$ where the superscript is suppressed.

We shall be interested in the situation where f is differentiable; namely, where each component function $f^i(t)$ is continuously differentiable in I. But this still does not give us a definition of a differentiable curve, for each $f^i(t)$ can be a constant so that the image of this mapping is a point. We, therefore, add the requirement below:

Definition. A trajectory is defined by a differentiable function $f\colon I \to R^n$ such that for each $t \in I$, $df^i(t)/dt$ is not a zero vector. The parameter t is called the parameter of the trajectory.

Definition. The vector $df^i(t)/dt$ is called a tangent vector. The set of tangent vectors corresponding to all $t \in I$ is called the tangent vector field.

The intuitive concept of a curve suggests that it should be independent of the parameter used to define it, and we wish to be able to change parameters to suit our convenience. To handle this idea, let $J\colon\ c \le u \le d$ be an interval of R^1, and suppose that we have a function $t = g(u)$ defined in J such that

(1) $g\colon J \to I$ is one-to-one, $g(c) = a$, $g(d) = b$.

(2) $dt/du > 0$ in J

Then we can write

$$\begin{aligned} x^i &= f^i(t) \\ &= f^i[g(u)] \\ &= h^i(u) \end{aligned}$$

Lemma 1. $x^i = h^i(u)$ is a trajectory. For $dh^i/du = \dfrac{[df^i(t)]}{dt}\dfrac{dt}{du}$ and hence exists and is a nonzero vector at each point of J.

Remark. Since $t = g(u)$ is increasing, the inverse function $u = g^{-1}(t)$ exists and is differentiable and increasing. So if $x^i = h^i(u)$ is a trajectory, so is $h^i[u(t)] = f^i(t)$. Hence two trajectories can be called equivalent if their parameters are related by a function satisfying the conditions above. It is easy to show that this is an equivalence relation. Now we can define a curve:

Definition. A differentiable, directed curve (or curve for short) in R^n is an equivalence class of trajectories. Any parameter corresponding to a trajectory in this class is called an allowable parameter.

4.3 Arc Length. The length L of an arc of a curve bounded by $t = t_1$, $t = t_2$ where $t_1 \leq t_2$ is defined to be

$$L = \int_{t_1}^{t_2} \left(\sum_i \left(\frac{df^i}{dt} \right)^2 \right)^{1/2} dt$$

Let

$$s = \int_a^t \left(\sum_i \left(\frac{df^i}{dt} \right)^2 \right)^{1/2} dt$$

Then

$$\frac{ds}{dt} = \left(\sum_i \left(\frac{df^i}{dt} \right)^2 \right)^{1/2} > 0$$

Hence s is an allowable parameter, and the parametric equations can be written

$$x^i = h^i(s)$$

The tangent vector dh^i/ds is a unit vector. For

$$\sum_i \left(\frac{dh^i}{ds} \right)^2 = \sum_i \left(\frac{df^i}{dt} \right)^2 \left(\frac{dt}{ds} \right)^2$$

$$= \frac{\sum_i \left(\frac{df^i}{dt}\right)^2}{\left(\frac{ds}{dt}\right)^2}$$

$$= \frac{\sum_i \left(\frac{df^i}{dt}\right)^2}{\sum_i \left(\frac{df^i}{dt}\right)^2}$$

$$= 1$$

We write $T = dh/ds$, where T is a vector, called the unit tangent vector.

4.4 <u>Frenet Equations of a Curve</u>. Before starting this topic, let us prove a lemma that we shall use repeatedly.

<u>Lemma 2</u>. If $V(t)$ is a field of unit vectors defined along a curve C in R^n, then dV/dt is normal to V provided that $dV/dt \neq 0$.

<u>Proof</u>. By hypothesis $\langle V,V \rangle = 1$. Hence

$$\langle dV/dt, V \rangle + \langle V, dV/dt \rangle = 0.$$

Since the inner product $\langle\ ,\ \rangle$ is symmetric, it follows that $\langle dV/dt, V \rangle = 0$ and so dV/dt is normal to V.

Now let $X = H(s)$ be the parametric equations of a curve C in R^n where s represents arc length. Then we have the unit tangent vector $T(s)$ given by

(1) $$\frac{dX}{ds} = T$$

Next consider dT/ds. We make the restrictive assumption that this is not zero, and so by Lemma 2 we can write

$$(2) \qquad \frac{dT}{ds} = K_1 N_1$$

where N_1 is a unit vector normal to T and $K_1 (\neq 0)$ is a factor of proportionality. N_1 is called the <u>first normal</u> vector to C and K_1 is the <u>first curvature</u> of C. Note that N_1 and K_1 are unique to within algebraic sign.

Next, we calculate dN_1/ds. By Lemma 2 this is normal to N_1. If we let aT be its projection on T, we can then write

$$\frac{dN_1}{ds} = aT + R$$

where R is normal to T and N_1. Now

$$a = \langle dN_1/ds, T \rangle = -\langle N_1, dT/ds \rangle = -K_1.$$

So if we assume $R \neq 0$, we may write

$$(3) \qquad \frac{dN_1}{ds} = -K_1 T + K_2 N_2$$

where N_2 (the <u>second normal</u>) is a unit vector orthogonal to T and N_1; and $K_2 (\neq 0)$ is the <u>second curvature</u> of C.

Continuing in this way we derive the equations:

$$(4) \qquad \frac{dN_2}{ds} = -K_2 N_1 + K_3 N_3$$

$$- - -$$

(n) $$\frac{dN_{n-2}}{ds} = -K_{n-2}N_{n-3} + K_{n-1}N_{n-1}$$

where $T, N_1, \ldots, N_{n-1}$ are an orthonormal set of unit vectors in R^n.

We complete this process with the last equation:

(n+1) $$\frac{dN_{n-1}}{ds} = -K_{n-1}N_{n-2}$$

In this derivation we had to assume that $K_1,\ldots,K_{n-2}$ were all nonzero in order to obtain the unit normal vectors $N_1,\ldots,N_{n-2}$. It is not necessary to assume $K_{n-1} \neq 0$ in order to define N_{n-1}, for we can define N_{n-1} as the unit vector orthogonal to $T, N_1,\ldots,N_{n-2}$ which had already been determined.

The system of equations (1) to (n+1) is called the Frenet equations of the curve C.

When $n = 3$, N_1 is called the <u>principal normal</u> and K_1, the <u>curvature</u>. Also N_2 is the <u>binormal</u> and K_2 is the <u>torsion</u>.

The Frenet system can be written more compactly as:

$$\begin{cases} \dfrac{dX}{ds} = N_o \\[2ex] \dfrac{dN_\alpha}{ds} = -K_\alpha N_{\alpha-1} + K_{\alpha+1}N_{\alpha+1} & \alpha = 0,1,\ldots,n-1 \end{cases}$$

where $N_o = T$, $K_o = 0$, and $K_n = 0$.

We then have Theorem 1.

<u>Theorem 1</u>. If $X = H(s)$ is a sufficiently differentiable curve C in R^n, there exist curvatures $K_1(s), \ldots, K_{n-1}(s)$ and an orthonormal frame $T(s), N_1(s), \ldots, N_{n-1}(s)$ which satisfy the Frenet equations provided that $K_1(s), \ldots, K_{n-2}(s)$ are not zero at any point of C. These quantities are unique to within algebraic sign.

4.5 <u>Solution of the Frenet Equations</u>. Here we consider the converse of Theorem 1. We assume that we have continuous functions $K_1(s), \ldots, K_{n-2}(s)$ all of which are nonzero, and $K_{n-1}(s)$ [which may have zeros or be identically zero]. Then we wish to solve the Frenet equations for $X(s)$, $T(s)$, $N_1(s), \ldots, N_{n-1}(s)$. But the problem is a bit more complicated, for we want T, N_1, $\ldots, N_{n-1}$ to form an orthonormal frame at each point of C. So we wish to solve the <u>mixed system</u> of differential and algebraic equations:

$$\begin{cases} \dfrac{dX}{ds} = N_0 & \langle N_\alpha, N_\beta \rangle - \delta_{\alpha\beta} = 0 \\[2ex] \dfrac{dN_\alpha}{ds} = -K_\alpha N_{\alpha-1} + K_{\alpha+1} N_{\alpha+1} & \end{cases}$$

where the notation is explained in Section (4.4).

First we can solve the differential equations with arbitrary initial values for X and N_α. Since we wish the algebraic equations to be satisfied as well, we should choose the initial values of N_α to satisfy these. So we choose:

$$X_0 \text{ arbitrary, } (N_\alpha)_0 \text{ satisfying } \langle (N_\alpha)_0, (N_\beta)_0 \rangle - \delta_{\alpha\beta} = 0$$

Thus we have a fixed point X_o in R^n and a fixed orthonormal frame at X_o. So we solve the differential equations with these initial values and obtain:

$$X(s),\ N_\alpha(s) \quad \text{at each point of} \quad C.$$

We still must prove that $\langle N_\alpha(s), N_\beta(s)\rangle - \delta_{\alpha\beta} = 0.$ To do so we consider the differential equations:

$$\frac{d}{ds}\{\langle N_\alpha, N_\beta\rangle - \delta_{\alpha\beta}\} = \left\langle \frac{dN_\alpha}{ds}; N_\beta \right\rangle + \left\langle N_\alpha, \frac{dN_\beta}{ds} \right\rangle$$

Since $N_\alpha(s)$ and $N_\beta(s)$ are now known to satisfy the Frenet equations, these become:

$$\frac{d}{ds}\{\langle N_\alpha, N_\beta\rangle - \delta_{\alpha\beta}\} =$$

$$-K_\alpha\langle N_{\alpha-1}, N_\beta\rangle + K_{\alpha+1}\langle N_{\alpha+1}, N_\beta\rangle - K_\beta\langle N_\alpha, N_{\beta-1}\rangle + K_{\beta+1}\langle N_\alpha, N_{\beta+1}\rangle$$

These can be rewritten:

$$\frac{d}{ds}\{\langle N_\alpha, N_\beta\rangle - \delta_{\alpha\beta}\} =$$

$$-K_\alpha\{\langle N_{\alpha-1}, N_\beta\rangle - \delta_{\alpha-1,\beta}\} + K_{\alpha+1}\{\langle N_{\alpha+1}, N_\beta\rangle - \delta_{\alpha+1,\beta}\}$$

$$-K_\beta\{\langle N_\alpha, N_{\beta-1}\rangle - \delta_{\alpha,\beta-1}\} + K_{\beta+1}\{\langle N_\alpha, N_{\beta+1}\rangle - \delta_{\alpha,\beta+1}\}$$

So we have a linear system of differential equations whose dependent variables are $\{\langle N_\alpha, N_\beta\rangle - \delta_{\alpha\beta}\}$ for which these variables are zero at the initial point s_o. Therefore, they are identically zero.

Thus we have proved Theorem 2.

__Theorem 2__. Given a set of nonzero continuous functions $K_1(s), \ldots, K_{n-2}(s)$ and a continuous function $K_{n-1}(s)$, there is a curve C in R^n with these functions as its curvatures. Moreover, C is unique for a given initial point X_o and initial normal frame $T, N_1, \ldots, N_{n-1}$ at X_o. Thus C is unique to within rigid motions in R^n.

__Remark__. The restriction that $K_1,\ldots,K_{n-2}$ be nonzero is unpleasant, for it excludes many interesting curves including straight lines. The partial removal of this restriction has been discussed by K. Nomizu: "On Frenet Equations for Curves of Class C^∞", Tohoku Math. Journal, Vol. 11 (1959), pp. 106-112.

4.6 __Hypersurfaces in R^n__. We shall define a hypersurface M as an equivalence class of $n-1$ dimensional charts.

__Definition__. An n-1 dimensional chart in R^n is an imbedded submanifold given by the parametric equations

$$x^i = g^i(u)$$

defined in an open ball $U \subset R^{n-1}$ and where the matrix $(\partial g/\partial u)$ is of rank $n-1$ in U.

Two such charts $x = g(u)$ and $x = h(v)$ defined respectively in U and V are equivalent if there is a differentiable homeomorphism (a "diffeomorphism") $U \to V$ given by $u = f(v)$ whose Jacobian is nonzero at every point of V and such that

$$x = g(u) = g[f(v)] = h(v)$$

From this it follows that

$$x = h(v) = h[f^{-1}(u)] = g(u)$$

Unfortunately, there is no preferred set of parameters in this case similar to arc length for curves. So we must choose an arbitrary set of parameters, and then check that our results are independent of this choice.

4.7 __Arc Length on a Hypersurface__. First we fix the notations. M will be defined by the parametric equations $X = G(u)$ or $x^i = g^i(u^\alpha)$ where $i = 1...n$, $\alpha = 1, ..., n-1$. The tangent vectors are

$$T_\alpha = \frac{\partial X}{\partial u^\alpha}$$

but these are no longer unit vectors. If we change parameters, we find that

$$T_\alpha = \frac{\partial X}{\partial u^\alpha} = \frac{\partial X}{\partial v^\beta}\frac{\partial v^\beta}{\partial u^\alpha} = \bar{T}_\beta \frac{\partial v^\beta}{\partial u^\alpha}$$

where $\bar{T}_\beta$ are the tangent vectors when the v's are chosen as parameters. This is merely a change of basis in the tangent space.

__Remark__. Here and hereafter we assume the Einstein summation convention to the effect that when the same index appears twice in a term, a summation on that index is implied. Thus

$$\text{for } \sum_\beta \bar{T}_\beta \frac{\partial v^\beta}{\partial u^\alpha} \text{ we write simply } \bar{T}_\beta \frac{\partial v^\beta}{\partial u^\alpha}.$$

Now consider a curve C on M given by the equations $u^\alpha = f^\alpha(t)$. As a curve in R^n its equations are

$$X = G[f(t)]$$

If the arc length of C is s, we know that

$$\left(\frac{ds}{dt}\right)^2 = \left\langle \frac{dX}{dt}, \frac{dX}{dt} \right\rangle$$

$$= \left\langle \frac{\partial X}{\partial u^\alpha} \frac{du^\alpha}{dt}, \frac{\partial X}{\partial u^\beta} \frac{\partial u^\beta}{dt} \right\rangle$$

$$= \left\langle \frac{\partial X}{\partial u^\alpha}, \frac{\partial X}{\partial u^\beta} \right\rangle \frac{du^\alpha}{dt} \frac{du^\beta}{dt}$$

We write $g_{\alpha\beta} = \left\langle \frac{\partial X}{\partial u^\alpha}, \frac{\partial X}{\partial u^\beta} \right\rangle = \left\langle T_\alpha, T_\beta \right\rangle = g_{\beta\alpha}$

So

$$\left(\frac{ds}{dt}\right)^2 = g_{\alpha\beta} \frac{du^\alpha}{dt} \frac{du^\beta}{dt}$$

Since the left-side is always positive, it follows that the matrix $(g_{\alpha\beta})$ is positive definite, and hence is nonsingular.

If we change parameters, we find that

$$g_{\alpha\beta} = \left\langle \overline{T}_\delta \frac{\partial v^\delta}{\partial u^\alpha}, \overline{T}_\varepsilon \frac{\partial v^\varepsilon}{\partial u^\beta} \right\rangle$$

$$= \left\langle \overline{T}_\delta, \overline{T}_\varepsilon \right\rangle \frac{\partial v^\delta}{\partial u^\alpha} \frac{\partial v^\varepsilon}{\partial u^\beta}$$

$$= \overline{g}_{\delta\varepsilon} \frac{\partial v^\delta}{\partial u^\alpha} \frac{\partial v^\varepsilon}{\partial u^\beta}$$

As we shall see in Chapter 5, this proves that $g_{\alpha\beta}$ are the components of a <u>tensor</u> of type $(0,2)$.

Thus $g_{\alpha\beta}$ is called the <u>first fundamental tensor</u> of M, and

$$ds^2 = g_{\alpha\beta}du^\alpha du^\beta$$

is called the <u>first fundamental form</u>. Often $g_{\alpha\beta}$ is called the metric tensor, for it appears in the formula for length of any curve on M. But clearly it is not a "metric" in the sense of metric in "metric space".

<u>Remark</u>. The great contribution of Riemann (1858) was to abstract these ideas from hypersurfaces to abstract differentiable manifolds. When such a tensor $g_{\alpha\beta}$ exists on an abstract manifold, it defines a Riemann metric; and thus we can begin the study of Riemannian geometry. As a variant of this, Einstein in his theory of General Relativity introduced a $g_{\alpha\beta}$ which was non-singular, but which was not positive definite. When his $g_{\alpha\beta}$ is diagonalized, its signs have the pattern $\begin{pmatrix} + & & \\ & - & \\ & & - \end{pmatrix}$.

4.8 <u>The Frenet Equations of a Hypersurface</u>. Since we are given n-1 independent vectors T_α, we can construct the unit normal N to the hypersurface so that $\{T_\alpha, N\}$ is a basis for R^n. Then Then $\partial T_\alpha/\partial u_\beta$ is a vector in R^n, and can be expressed in terms of this basis in the following way:

$$\frac{\partial T_\alpha}{\partial u^\beta} = \Gamma^{\gamma}_{\alpha\beta} T_\gamma + b_{\alpha\beta}N \tag{1}$$

where $\Gamma^{\gamma}_{\alpha\beta}$ and $b_{\alpha\beta}$ are coefficients of linear dependence.

From (1) we find that

$$b_{\alpha\beta} = \langle \partial T_\alpha/\partial u^\beta, N\rangle = \langle \partial T_\beta/\partial u^\alpha, N\rangle = b_{\beta\alpha}$$

These are also components of a tensor of type (0,2), called the <u>second fundamental tensor</u> of M. If we compare equation (1) above with equation (2) of section (4.4), we see that the matrix $(b_{\alpha\beta})$ is a generalization of the first curvature of a curve. The expression

$$b_{\alpha\beta}dx^{\alpha}dx^{\beta}$$

is called the <u>second fundamental form</u> of M.

For the moment we pass over the needed discussion of $\Gamma^{\gamma}_{\alpha\beta}$, and proceed with the construction of the Frenet system. We must consider $\partial N/\partial u^{\alpha}$. We find (using Lemma 2):

$$\frac{\partial N}{\partial u^{\alpha}} = C^{\beta}_{\alpha}\, T_{\beta}$$

where C^{β}_{α} are coefficients of linear dependence. From this, we derive:

$$\langle \partial N/\partial u_{\alpha}, T_{\gamma} \rangle = C^{\beta}_{\alpha} \langle T_{\beta}, T_{\gamma} \rangle = C^{\beta}_{\alpha}\, g_{\beta\gamma}$$

But

$$\langle \partial N/\partial u_{\alpha}, T_{\gamma} \rangle = - \langle N, \partial T_{\gamma} / \partial u^{\alpha} \rangle = -b_{\alpha\gamma}$$

So

$$C^{\beta}_{\alpha}\, g_{\beta\gamma} = -b_{\alpha\gamma}$$

Let $g^{\beta\gamma}$ be the matrix inverse to $g_{\beta\gamma}$. Then we have

$$C^{\beta}_{\alpha} = -g^{\beta\gamma}b_{\alpha\gamma}$$

Hence, we finally obtain

(2) $$\frac{\partial N}{\partial u^\alpha} = -g^{\beta\gamma} b_{\alpha\gamma} T_\beta \ .$$

In summary, the Frenet system becomes

$$\begin{cases} \dfrac{\partial X}{\partial u^\alpha} = T_\alpha \\[2ex] \dfrac{\partial T^\alpha}{\partial u^\beta} = \Gamma^\gamma_{\alpha\beta} T_\gamma + b_{\alpha\beta} N \\[2ex] \dfrac{\partial N}{\partial u^\alpha} = -g^{\beta\gamma} b_{\alpha\gamma} T_\beta \end{cases}$$

together with the algebraic conditions

$$\langle T_\alpha, T_\beta \rangle - g_{\alpha\beta} = 0$$

$$\langle T_\alpha, N \rangle = 0$$

$$\langle N, N \rangle - 1 = 0$$

As in the case of curves, these equations imply all the differential geometry of our hypersurface.

4.9 <u>The Coefficients $\Gamma^\gamma_{\alpha\beta}$</u>. This is the first appearance of what will later be called a <u>connection</u>, which is one of the most important structures on a differentiable manifold. The definition of a connection there is an abstraction from the properties of $\Gamma^\gamma_{\alpha\beta}$ that we shall develop in this chapter.

The first important property is given in Theorem 3.

<u>Theorem 3</u>. $\Gamma^{\gamma}_{\alpha\beta} = g^{\gamma\delta}[\alpha\beta,\delta]$

where $[\alpha\beta,\delta] = 1/2\left(\dfrac{\partial g_{\alpha\delta}}{\partial u^{\beta}} + \dfrac{\partial g_{\beta\delta}}{\partial u^{\alpha}} - \dfrac{\partial g_{\alpha\beta}}{\partial u^{\delta}}\right)$

The symbols $[\alpha\beta,\delta]$ and $\Gamma^{\gamma}_{\alpha\beta}$ are called, respectively, the Christoffel symbols of the <u>first</u> and <u>second</u> <u>kinds</u>.

<u>Proof</u>. For convenience we write $T_{\alpha\beta} = T_{\beta\alpha} = \partial T_{\alpha}/\partial u^{\beta}$.

Then
$$\frac{\partial}{\partial u^{\gamma}}(g_{\alpha\beta}) = \frac{\partial}{\partial u^{\gamma}}\langle T_{\alpha},T_{\beta}\rangle$$
$$= \langle T_{\alpha\gamma},T_{\beta}\rangle + \langle T_{\alpha},T_{\beta\gamma}\rangle$$

So by their definition:

$$[\alpha\beta,\delta] = 1/2\Big(\langle T_{\alpha\beta},T_{\delta}\rangle + \langle T_{\alpha},T_{\delta\beta}\rangle + \langle T_{\beta\alpha},T_{\delta}\rangle + \langle T_{\beta},T_{\delta\alpha}\rangle$$
$$- \langle T_{\alpha\delta},T_{\beta}\rangle - \langle T_{\alpha},T_{\beta\delta}\rangle\Big)$$
$$= \langle T_{\alpha\beta},T_{\delta}\rangle$$

So from equation (1) of Section 4.8 it follows that

$$[\alpha\beta,\delta] = \Gamma^{\gamma}_{\alpha\beta}\langle T_{\gamma},T_{\delta}\rangle = \Gamma^{\gamma}_{\alpha\beta}\, g_{\gamma\delta}$$

Hence $\quad \Gamma^{\gamma}_{\alpha\beta} = g^{\gamma\delta}[\alpha\beta,\delta]$

The importance of this result is that $\Gamma^{\gamma}_{\alpha\beta}$ can be calculated from a knowledge of $g_{\alpha\beta}$ and its first derivatives.

The second important property of $\Gamma^{\gamma}_{\alpha\beta}$ is that when we change parameters, these expressions turn out not to be tensors. They are something quite different, and thereby hangs a tale.

In order to show this, we recall that

$$T_\alpha = \overline{T}_\beta \frac{\partial v^\beta}{\partial u^\alpha}$$

So

$$T_{\alpha\delta} = \overline{T}_{\beta\varepsilon} \frac{\partial v^\beta}{\partial u^\alpha} \frac{\partial v^\varepsilon}{\partial u^\delta} + \overline{T}_\beta \frac{\partial^2 v^\beta}{\partial u^\alpha \partial u^\delta}$$

To keep our formulas in manageable shape we shall hereafter write

$$A^\beta_\alpha = \frac{\partial v^\beta}{\partial u^\alpha} \; ; \; \left(A^\alpha_\beta\right)^{-1} = \frac{\partial u^\alpha}{\partial v^\beta} \; ; \; A^\beta_{\alpha\delta} = \frac{\partial^2 v^\beta}{\partial u^\alpha \partial u^\delta}$$

Then

$$T_\alpha = \overline{T}_\beta A^\beta_\alpha \; ; \; T_{\alpha\delta} = \overline{T}_{\beta\varepsilon} A^\beta_\alpha A^\varepsilon_\delta + \overline{T}_\beta A^\beta_{\alpha\delta}$$

In this notation a set of quantities $Q^{\alpha_1 \ldots \alpha_r}_{\beta_1 \ldots \beta_s}$ are the components of a tensor of type (r,s) if their law of transformation is

$$\overline{Q}^{\,\nu_1 \ldots \nu_r}_{\,\delta_1 \ldots \delta_s} = Q^{\alpha_1 \ldots \alpha_r}_{\beta_1 \ldots \beta_s} A^{\nu_1}_{\alpha_r} \ldots A^{\nu_r}_{\alpha_r} (A^{-1})^{\beta_1}_{\delta_1} \ldots (A^{-1})^{\beta_s}_{\nu_s}$$

Thus, in this notation

$$[\alpha\delta, \gamma] = \langle T_{\alpha\delta}, T_\gamma \rangle$$

$$= \langle \overline{T}_{\beta\varepsilon} A^\beta_\alpha A^\varepsilon_\delta + \overline{T}_\beta A^\beta_{\alpha\delta}, \; \overline{T}_\lambda A^\lambda_\gamma \rangle$$

$$= \overline{[\beta\varepsilon,\lambda]}\, A^{\beta}_{\alpha}A^{\varepsilon}_{\delta}A^{\lambda}_{\gamma} + \bar{g}_{\beta\lambda}A^{\beta}_{\alpha\delta}A^{\lambda}_{\gamma}$$

The presence of the second derivatives $\left(A^{\beta}_{\alpha\delta}\right)$ shows that $[\alpha\delta,\gamma]$ is not a tensor.

From the fact that

$$g^{\gamma\sigma} = \bar{g}^{\rho\tau}\left(A^{\gamma}_{\rho}\right)^{-1}\left(A^{\sigma}_{\tau}\right)^{-1}$$

it can be proved (after a messy calculation) that

$$(1) \quad \Gamma^{\sigma}_{\alpha\delta} = \bar{\Gamma}^{\tau}_{\beta\varepsilon}A^{\beta}_{\alpha}A^{\varepsilon}_{\delta}\left(A^{\sigma}_{\tau}\right)^{-1} + A^{\tau}_{\alpha\delta}\left(A^{\sigma}_{\tau}\right)^{-1}$$

Thus the $\Gamma^{\sigma}_{\alpha\delta}$ are not components of a tensor.

Equation (1) may also be written in the useful form:

$$(2) \quad \Gamma^{\sigma}_{\alpha\delta}A^{\tau}_{\sigma} = \bar{\Gamma}^{\tau}_{\beta\varepsilon}A^{\beta}_{\alpha}A^{\varepsilon}_{\delta} + A^{\tau}_{\alpha\delta}$$

Equations (1) and (2) have been called the fundamental equations of Riemannian geometry. We have derived them on the assumption that M is a hypersurface of R^n. But if we have a Riemann metric $g_{\alpha\beta}$ in any manifold whatever and define $\Gamma^{\gamma}_{\alpha\beta}$ by the formulas of Theorem 3, equations (1) and (2) are true. In this more general setting they can be proved by a direct (but very messy) calculation, or by the use of advanced theorems. For lack of time and space, we shall not pursue this farther.

4.10 Covariant Derivatives of Vectors. We have seen that T_{α} form a basis for the tangent space to M at a fixed point. Thus any tangent vector, T, has an expression of the form:

$$T = \lambda^{\alpha} T_{\alpha}$$

T is a vector in R^n which makes sense to an observer in R^n, but it makes no sense to an observer who is restricted to M. The intrinsic observer in M can only make sense of λ^{α}, and so it is natural for him to call λ^{α} the components of a vector. Let us see how they transform under changes of parameter. We must have

$$T = \lambda^{\alpha} T_{\alpha} = \bar{\lambda}^{\beta} \bar{T}_{\beta}$$

But

$$T_{\alpha} = \bar{T}_{\beta} A^{\beta}_{\alpha}$$

So

$$\lambda_{\alpha} T_{\alpha} = \left(\lambda^{\alpha} A^{\beta}_{\alpha}\right) \bar{T}_{\beta} = \bar{\lambda}^{\beta} \bar{T}_{\beta}$$

Since $\bar{T}_{\beta}$ are independent, we conclude that

(1) $$\bar{\lambda}^{\beta} = \lambda^{\alpha} A^{\beta}_{\alpha}$$

These are just the equations for the transformation of the components of a vector under a change of basis in a vector space. In tensor terminology λ^{α} are the components of a tensor of type (1,0).

Next consider the derivative of such a vector field. From (1)

$$\frac{\partial \bar{\lambda}^{\beta}}{\partial v^{\delta}} = \frac{\partial \lambda^{\alpha}}{\partial u^{\gamma}} \left(A^{\gamma}_{\delta}\right)^{-1} A^{\beta}_{\alpha} + \lambda^{\alpha} A^{\beta}_{\alpha\gamma} \left(A^{\gamma}_{\delta}\right)^{-1}$$

or $$\frac{\partial \bar{\lambda}^{\beta}}{\partial v^{\delta}} A^{\delta}_{\gamma} = \frac{\partial \lambda^{\alpha}}{\partial u^{\gamma}} A^{\beta}_{\alpha} + \lambda^{\alpha} A^{\beta}_{\alpha\gamma}$$

From equation (2) of section 4.9, we know that

$$A^{\beta}_{\alpha\gamma} = \Gamma^{\varepsilon}_{\alpha\gamma} A^{\beta}_{\varepsilon} - \bar{\Gamma}^{\beta}_{\varepsilon\delta} A^{\varepsilon}_{\alpha} A^{\delta}_{\gamma}$$

So

$$\frac{\partial \bar{\lambda}^{\beta}}{\partial v^{\delta}} A^{\delta}_{\gamma} = \frac{\partial \lambda^{\alpha}}{\partial u^{\gamma}} A^{\beta}_{\alpha} + \lambda^{\alpha} \Gamma^{\varepsilon}_{\alpha\gamma} A^{\beta}_{\varepsilon} - \lambda^{\alpha} \bar{\Gamma}^{\beta}_{\varepsilon\delta} A^{\varepsilon}_{\alpha} A^{\delta}_{\gamma}$$

$$= \frac{\partial \lambda^{\alpha}}{\partial u^{\gamma}} A^{\beta}_{\alpha} + \lambda^{\delta} \Gamma^{\alpha}_{\delta\gamma} A^{\beta}_{\alpha} - \bar{\lambda}^{\varepsilon} \bar{\Gamma}^{\beta}_{\varepsilon\delta} A^{\delta}_{\gamma}$$

Thus:

$$\left(\frac{\partial \bar{\lambda}^{\beta}}{\partial v^{\delta}} + \bar{\lambda}^{\varepsilon} \bar{\Gamma}^{\beta}_{\varepsilon\delta}\right) A^{\delta}_{\gamma} = \left(\frac{\partial \lambda^{\alpha}}{\partial u^{\gamma}} + \lambda^{\delta} \Gamma^{\alpha}_{\delta\gamma}\right) A^{\beta}_{\alpha}$$

If we write

$$\lambda^{\alpha}_{,\gamma} = \frac{\partial \lambda^{\alpha}}{\partial u^{\gamma}} + \lambda^{\delta} \Gamma^{\alpha}_{\delta\gamma}$$

this becomes

$$\bar{\lambda}^{\beta}_{,\delta} = \lambda^{\alpha}_{,\gamma} A^{\beta}_{\alpha} \left(A^{\gamma}_{\delta}\right)^{-1}$$

which shows that $\lambda^{\alpha}_{,\gamma}$ have a tensor law of transformation and are the components of a tensor of type (1,1). This tensor is called the covariant derivative of λ^{α}.

Related to the covariant derivative is the directional covariant derivative

$$\lambda^{\alpha}_{,\gamma} \xi^{\gamma}$$

where ξ^γ is another vector field. We prove the lemma:

Lemma 3. The directional covariant derivative of λ^α in the direction of ξ^γ is a vector field.

Proof. We know that

$$\bar{\lambda}^\beta_{,\delta} = \lambda^\alpha_{,\gamma} A^\beta_\alpha (A^\gamma_\delta)^{-1} \quad \text{and} \quad \bar{\xi}^\delta = \xi^\varepsilon A^\delta_\varepsilon$$

So

$$\bar{\lambda}^\beta_{,\delta} \bar{\xi}^\delta = \lambda^\alpha_{,\gamma} A^\beta_\alpha (A^\gamma_\delta)^{-1} \xi^\varepsilon A^\delta_\varepsilon$$

$$= \left(\lambda^\alpha_{,\gamma} \xi^\gamma\right) A^\beta_\alpha$$

This shows that $\lambda^\alpha_{,\gamma} \xi^\gamma$ are components of a vector.

In the modern abstract definition of a connection to come later, we choose intrinsic tangent vectors X, Y, Z with components X^α, Y^β, Z^γ respectively. Then we define the operator ∇_X as follows:

(1) $\nabla_X f = (df)(X)$ where f is a function

(2) $\nabla_X Y = Y^\alpha_{,\gamma} X^\gamma$, a vector field.

From these definitions we derive the properties (proofs left as exercises):

(3) $\nabla_X (Y + Z) = \nabla_X Y + \nabla_X Z$

(4) $\nabla_{X+Y}(Z) = \nabla_X Z + \nabla_Y Z$

(5) $\nabla_{fX}(Y) = f\,\nabla_X Y$

(6) $\nabla_X(fY) = (\nabla_X f)\,Y + f\nabla_X Y$

It can be shown that properties (1) to (6) characterize the covariant derivative. It has been our purpose here to derive these properties as motivation for our later discussion of connections on abstract manifolds, and to give a concrete example to show that the abstract theory is not vacuous. So the further discussion of this topic is deferred.

4.11 The Principal Curvatures of a Hypersurface. Let C be a curve on a hypersurface M defined by $u^\alpha = f^\alpha(s)$ or by $X(s) = X[u(s)]$ where s is the arc length of C. Then T, the tangent to C, is given by:

$$T = \frac{dX}{ds} = \frac{\partial X}{\partial u^\alpha}\frac{du^\alpha}{ds} = T_\alpha \frac{du^\alpha}{ds}$$

and

$$\frac{dT}{ds} = \frac{d^2X}{ds^2} = T_{\alpha\beta}\frac{du^\alpha}{ds}\frac{du^\beta}{ds} + T_\alpha \frac{d^2u^\alpha}{ds^2}$$

$$= \left(\Gamma^\gamma_{\alpha\beta}\frac{du^\alpha}{ds}\frac{du^\beta}{ds}T_\gamma + b_{\alpha\beta}\frac{du^\alpha}{ds}\frac{du^\beta}{ds}N\right) + T_\alpha\frac{d^2u^\alpha}{ds^2}$$

Thus

$$K_c N_c = \left(\frac{d^2u^\gamma}{ds^2} + \Gamma^\gamma_{\alpha\beta}\frac{du^\alpha}{ds}\frac{du^\beta}{ds}\right)T_\gamma + b_{\alpha\beta}\frac{du^\alpha}{ds}\frac{du^\beta}{ds}N$$

where N_c is the first normal to C and K_c is its first curvature.

Define

$$G^{\gamma} = \frac{d^2u^{\gamma}}{ds^2} + \Gamma^{\gamma}_{\alpha\beta}\frac{du^{\alpha}}{ds}\frac{du^{\beta}}{ds}$$

$$(K_g)^2 = g_{\alpha\beta}\, G^{\alpha}G^{\beta}$$

Then K_g is the <u>geodesic curvature</u> of C.

Also define

$$K_N = b_{\alpha\beta}\frac{du^{\alpha}}{ds}\frac{du^{\beta}}{ds}$$

where now K_N is the <u>normal curvature</u> of C. Then we have the theorem:

<u>Theorem 4</u>. $K_c{}^2 = K_g{}^2 + K_N{}^2$

<u>Definition</u>. C is a geodesic iff $N_c = \pm N$; or iff $G^k = 0$ for all s.

<u>Remark</u>. Geodesics also have extremal properties that we shall not discuss here.

To find the geometrical interpretation of the second fundamental form, consider a fixed point P on M. At P consider the unit tangent vector $\Lambda = T_{\alpha}\lambda^{\alpha}$ where λ^{α} are real numbers. Since Λ is a unit vector

$$1 = \langle \Lambda, \Lambda \rangle = g_{\alpha\beta}\, \lambda^{\alpha}\lambda^{\beta}$$

We call Λ a direction on M.

Let C be a curve on M through P having the properties:

(1) C is tangent to the chosen direction at P;

(2) The geodesic curvature of C at P is zero.

There are many examples of such curves C, but two examples are as follows:

(a) C is a geodesic through P. We obtain this by integrating the differential equations:

$$\frac{d^2u^\alpha}{ds^2} + \Gamma^\alpha_{\beta\delta}\frac{du^\beta}{ds}\frac{du^\delta}{ds} = 0$$

with initial conditions at $s = 0$: $u^\alpha(0) = u^\alpha(P)$; $\left(\frac{du^\alpha}{ds}\right)_0 = \lambda^\alpha$.

(b) Let π be a plane through N (at P) and Λ. Then π intersects M in a plane curve C for which $N_c = \pm N$.

For any of these choices we can choose N_c such that $N_c = +N$ and $K_c = b_{\alpha\beta}\lambda^\alpha\lambda^\beta$. As Λ takes all possible positions, K_c will vary and for certain Λ will assume extreme values. These are obtained by the method of Lagrange multipliers where we seek extreme values of $b_{\alpha\beta}\lambda^\alpha\lambda^\beta$ subject to the condition that $g_{\alpha\beta}\lambda^\alpha\lambda^\beta = 1$. Hence we must solve:

$$\begin{cases} (b_{\alpha\beta} - \mu g_{\alpha\beta})\lambda^\beta = 0 & \text{or} \quad (B - \mu G)\Lambda = 0 \\ g_{\alpha\beta}\lambda^\alpha\lambda^\beta = 1 & \qquad \Lambda^T G \Lambda = 1 \end{cases}$$

for μ and λ^β. For a solution to exist μ must satisfy

$$\det (b_{\alpha\beta} - \mu g_{\alpha\beta}) = 0 \qquad \text{or} \qquad |B - \mu G| = 0$$

which is a polynomial of degree n-1 in μ, and has n-1 roots. At this point we state two theorems of Linear Algebra.

<u>Lemma 4</u>. The roots of $|B - \mu G| = 0$ are all real, since G is positive definite and B is symmetric.

Lemma 5. If $\mu_1 \neq \mu_2$, the corresponding characteristic vectors Λ_1 and Λ_2 are orthogonal.

The corresponding directions thus form an orthonormal set of principal directions. The corresponding

$$K_\alpha = \mu_\alpha = b_{\beta\delta}\lambda^\beta_\alpha\lambda^\delta_\alpha$$

are called the principal curvatures of M at P.

Given the principal curvatures $K_1 \ldots K_{n-1}$, we form their elementary symmetric functions:

$$K_1 + K_2 + \ldots + K_{n-1}$$

$$K_1K_2 + K_1K_3 + \ldots + K_{n-2}K_{n-1}$$

$$K_1K_2 \ldots K_{n-1}$$

In the case of $M^2 \subset R^3$, Gauss called

$$H = \frac{K_1 + K_2}{2}$$

the mean curvature of M at P and

$$K = K_1K_2$$

the total curvature of M at P.

In the general case we call

$$K_T = K_1K_2 \ldots K_{n-1} = \frac{\det (b_{\alpha\beta})}{\det (g_{\alpha\beta})}$$

the Gaussian curvature of M at P. For n-1 even, K_T does not depend on the choice of sign in the determination of N.

The other symmetric functions are called the <u>mean curvatures</u> of M at P. They have expressions (a bit complicated) in terms of $b_{\alpha\beta}$ and $g_{\alpha\beta}$. The most important is:

$$K_1 + \ldots + K_{n-1} = g^{\alpha\beta} b_{\alpha\beta}$$

which is often called <u>the</u> mean curvature.

4.12 <u>The Gauss and Codazzi Equations</u>. If we return to the differential equations of the Frenet system of Section 4.8, we naturally ask about their integrability supposing that $g_{\alpha\beta}(u)$ and $b_{\alpha\beta}(u)$ are known functions. By the usual process we can compute their integrability conditions, and find that they reduce to the following:

<u>Gauss</u>. $R_{\alpha\beta\gamma\delta} = b_{\alpha\gamma} b_{\beta\delta} - b_{\alpha\delta} b_{\beta\gamma}$

where $$R_{\alpha\beta\gamma\delta} = g_{\beta\varepsilon} \left[\frac{\partial \Gamma^{\varepsilon}_{\alpha\gamma}}{\partial u^{\delta}} - \frac{\partial \Gamma^{\varepsilon}_{\alpha\delta}}{\partial u^{\gamma}} + \Gamma^{\lambda}_{\alpha\gamma} \Gamma^{\varepsilon}_{\lambda\delta} - \Gamma^{\lambda}_{\alpha\delta} \Gamma^{\varepsilon}_{\lambda\gamma} \right]$$

<u>Codazzi</u>. $b_{\alpha\beta,\delta} = b_{\alpha\delta,\beta}$

where $$b_{\alpha\beta,\delta} = \frac{\partial b_{\alpha\beta}}{\partial u^{\delta}} - \Gamma^{\varepsilon}_{\alpha\delta} b_{\varepsilon\beta} - \Gamma^{\varepsilon}_{\beta\delta} b_{\varepsilon\alpha}$$

Since $R_{\alpha\beta\gamma\delta}$ are expressed in terms of $g_{\alpha\beta}$ and their derivatives, it follows from the Gauss equations that the 2×2 minors of $b_{\alpha\beta}$ depend in the same way on $g_{\alpha\beta}$. They are, therefore, called <u>intrinsic</u>. This means that they can be computed by an observer whose vision is restricted to the hypersurface even though to compute the individual components of $b_{\alpha\beta}$ an observer would have to live in R^n. We then have the celebrated "Theorema Egregium" of Gauss:

Theorem 5. For $n-1$ even, the total curvature K_T of M at every point depends only on the $g_{\alpha\beta}$ and their derivatives. That is, K_T is intrinsic.

Proof. $K_T = \dfrac{\det(b_{\alpha\beta})}{\det(g_{\alpha\beta})}$

And since the dimension $n-1$ is even, $\det(b_{\alpha\beta})$ can be expressed as a polynomial in its 2×2 minors.

Remark. This theorem is false if $n-1$ is odd, for then $\det(b_{\alpha\beta})$ cannot be expressed in terms of its 2×2 minors.

We then face the following problem. Suppose that on an open set $U \subset R^{n-1}$ we are given a positive definite matrix $(g_{\alpha\beta})$ of C^3 functions and a symmetric matrix $(b_{\alpha\beta})$ of C^2 functions which satisfy the Gauss and Codazzi equations. Can we solve the mixed system:

$$\begin{cases} \dfrac{\partial X}{\partial u^\alpha} = T_\alpha \\ \dfrac{\partial T_\alpha}{\partial u^\beta} = \Gamma_{\alpha\beta}^{\gamma} T_\gamma + b_{\alpha\beta} N \\ \dfrac{\partial N}{\partial u^\alpha} = -b_{\alpha\beta} g^{\beta\gamma} T_\gamma \end{cases}$$

$$\begin{cases} \langle T_\alpha, T_\beta \rangle - g_{\alpha\beta} = 0 \\ \langle T_\alpha, N \rangle = 0 \\ \langle N, N \rangle - 1 = 0 \end{cases}$$

for the vectors: X, T_α, and N? As for curves in Section 4.5 we choose X_0 arbitrary, and $(T_\alpha)_0$ and N_0 satisfying the algebraic equations of the system. Then we can solve the differential equations with these initial values and obtain $X(u)$, $T_\alpha(u)$, and $N(u)$ in our open set U. To show that these satisfy the algebraic equations in U, we proceed as in Section 4.5 to obtain a system of linear partial differential equations satisfied by $\{<T_\alpha, T_\beta> - g_{\alpha\beta}\}$, $<T_\alpha, N>$, and $\{<N,N> - 1\}$. Since these expressions are zero at the initial point by our choice of initial values, they are zero everywhere in U. We omit the rather complex details which add nothing to the ideas involved. Thus we have the result:

Theorem 6. If in $U \subset R^{n-1}$ we are given $g_{\alpha\beta}$ (positive definite) of class C^3 and $b_{\alpha\beta}$ (symmetric) of class C^2 which satisfy the Gauss and Codazzi equations, there exists a local hypersurface $M^{n-1} \subset R^n$ for which $g_{\alpha\beta}$ and $b_{\alpha\beta}$ are the first and second fundamental tensors respectively. Moreover, this hypersurface is unique to within motions in R^n.

4.13 Volume (area) of a Hypersurface. As a generalization of the cross product of two vectors in R^3, we consider the symbolic determinant:

$$\begin{vmatrix} e_1 & --- & e_n \\ T_1^1 & --- & T_1^n \\ --- & --- & --- \\ T_{n-1}^1 & --- & T_{n-1}^n \end{vmatrix}$$

where e_i are the usual basis of R^n and T_α^i are the R^n components of T_α. The symbolic minor of e_1 is

$$\begin{vmatrix} T_1^2 & --- & T_1^n \\ \cdot & & \cdot \\ \cdot & & \cdot \\ T_{n-1}^2 & --- & T_{n-1}^n \end{vmatrix} = \frac{\partial(x^2,\ldots,x^n)}{\partial(u^1,\ldots,u^{n-1})}$$

which hereafter we call V^1 for short. If we do this for each e_i in turn we get $V^2 \ldots V^n$ where

$$V^i = \frac{\partial(x^1,\ldots,\widehat{x^i}\ldots x^n)}{\partial(u^1,\ldots,u^{n-1})} \times (-1)^{i+1}$$

where $\widehat{x^i}$ means that x^i is to be omitted. These form a vector field in R^n: $V = (V^1,\ldots,V^n)$, which is the $(n-1)$ - fold "cross product" of $T_1,\ldots,T_{n-1}$. As in the case of R^3, V is orthogonal to each of $T_1,\ldots,T_{n-1}$. (Proof as in R^3). So V is a normal vector to M^{n-1}.

We must find an expression for the length of V. To do so, consider the determinant:

$$\Delta = \begin{vmatrix} V^1 & --- & V^n \\ T_1^1 & --- & T_1^n \\ \cdot & & \cdot \\ \cdot & & \cdot \\ T_{n-1}^1 & --- & T_{n-1}^n \end{vmatrix} \quad \text{which we write as} \quad \begin{vmatrix} V \\ T_1 \\ \cdot \\ \cdot \\ T_{n-1} \end{vmatrix}$$

From our definition of V, it follows that $\Delta = |V|^2$

Also

$$\Delta^2 = \Delta\Delta^T = \begin{vmatrix} V \\ T_1 \\ \cdot \\ \cdot \\ \cdot \\ T_{n-1} \end{vmatrix} \begin{vmatrix} V & T_1 & --- & T_{n-1} \end{vmatrix} = \begin{vmatrix} |V|^2 & 0 \\ 0 & g_{\alpha\beta} \end{vmatrix}$$

So $|V|^4 = |V|^2 \det(g_{\alpha\beta})$ or $|V| = \sqrt{\det(g_{\alpha\beta})} = \sqrt{g}$

where $g = \det(g_{\alpha\beta})$. Thus we have the result

<u>Theorem 7</u>. If $V^i = \dfrac{(x^1 \ldots \widehat{x^i} \ldots x^n)}{(u^1 \ldots u^{n-1})} \times (-1)^{i+1}$,

then V is a normal vector of length $\sqrt{g}$, or $V = \pm\sqrt{g}\, N$ depending on the choice of N. By convention we choose N so that $V = +\sqrt{g}\, N$.

Now we proceed to motivate the definition of volume on M^{n-1}. Choose a point $P(u^1, \ldots, u^{n-1}) \varepsilon M^{n-1}$, tangent vectors T_α at P and the normal N at P. Near P choose points of M^{n-1}

$$P_1 = (u^1 + \Delta u^1, u^2, \ldots, u^{n-1})$$
$$- - -$$
$$P_{n-1} = (u^1, u^2, \ldots, u^{n-1} + \Delta u^{n-1})$$

The points $P, P_1, \ldots, P_{n-1}$ are the vertices of an $(n-1)$ dimensional curvilinear parallelepiped whose area we wish to approximate. To do so we consider the points in R^{n-1}

$$\overline{P}_1 = P + T_1 \Delta u^1; \; \ldots, \; \overline{P}_{n-1} = P + T_{n-1} \Delta u^{n-1}$$

which are near $P_1, \ldots, P_{n-1}$ respectively. The points $P, \overline{P}_1, \ldots, \overline{P}_{n-1}$ are the vertices of a Euclidean parallelepiped whose volume is

$$\begin{vmatrix} N \\ T_1 \\ \cdot \\ \cdot \\ T_{n-1} \end{vmatrix} \Delta u^1 \times \dots \times \Delta u^{n-1} = \langle N, V \rangle \Delta u^1 \dots \Delta u^{n-1}$$

$$= \sqrt{g}\ \Delta u^1 \dots \Delta u^{n-1}.$$

Hence the following definition is reasonable.

Definition. The volume of a region R on M^{n-1} is defined to be

$$\int_R \sqrt{g}\ du^1 \dots du^{n-1}$$

4.14 The Spherical Image of M^{n-1} in R^n. Let us construct the "Gauss map" $M^{n-1} \to S^{n-1}$ where S^{n-1} is a unit sphere in R^n. The map is defined by: $X(u) \to N(u)$, where now $N(u)$ is the position vector of a point on S^{n-1}. We wish to find the element of area, $|\Delta S|$, of the image of this map.

From Section 4.13 we see that $|\Delta S|$ is the absolute value of:

$$\Delta S = \begin{vmatrix} N^1 & --- & N^n \\ \frac{\partial N^1}{\partial u^1} & --- & \frac{\partial N^n}{\partial u^1} \\ \cdot & & \cdot \\ \cdot & & \cdot \\ \cdot & & \cdot \\ \frac{\partial N^1}{\partial u^{n-1}} & --- & \frac{\partial N^n}{\partial u^{n-1}} \end{vmatrix} \Delta u^1 \dots \Delta u^{n-1}$$

To compute ΔS we observe that

$$\begin{vmatrix} N \\ \frac{\partial N}{\partial u^1} \\ \cdot \\ \cdot \\ \frac{\partial N}{\partial u^{n-1}} \end{vmatrix} \left| N\ T_1 \dots T_{n-1} \right| = \begin{vmatrix} 1 & 0 \\ 0 & -b_{\beta\gamma} \end{vmatrix} = (-1)^{n-1} \det(b_{\alpha\beta})$$

since $\left\langle \frac{\partial N}{\partial u^\alpha}, T_\beta \right\rangle = \left\langle -b_{\alpha\gamma} g^{\gamma\delta} T_\delta, T_\beta \right\rangle = -b_{\alpha\gamma} g^{\gamma\delta} g_{\beta\delta} = -b_{\alpha\beta}$

Therefore

$$|\Delta S| = |(\det b_{\alpha\beta}/\sqrt{g}\,|\Delta u^1 \ldots \Delta u^{n-1}$$

$$= |K_T|\sqrt{g}\,\Delta u^1 \ldots \Delta u^{n-1} = |K_T|\Delta V$$

where K_T is the Gaussian curvature of M^{n-1}, and ΔV is the volume of a patch on M^{n-1} whose image has volume $|\Delta S|$.

Thus we have the result:

Theorem 8. $|K_T|$ at a point P of M^{n-1} is given by

$$|K_T|_P = \lim_{\Delta V \to 0} \frac{|\Delta S|}{\Delta V}$$

The sign of K_T is that of the orientation of the image of the Gauss map. For a discussion of orientation, see Section 6.16 and exercise 21, Chapter 6.

Exercises

1. In R^3 find expressions for K_1 and K_2 if the curve is defined in terms of an arbitrary parameter by the equations $x^i = x^i(t)$; $i = 1,2,3$. Choose $N_2 = T_\wedge N_1$.

<u>Ans</u>. Let $X(t) = [x^1(t), x^2(t), x^3(t)]$
Then

$$|K_1| = \frac{|X'_\wedge X''|}{|X'|^3}; \quad K_2 = \frac{(X'_\wedge X'')\cdot X'''}{|X'_\wedge X''|^2}$$

where $_\wedge$ indicates the vector cross product.

2. As in Problem 1 find expressions for T, N_1, and N_2

<u>Ans</u>. $T = \dfrac{X'}{|X'|}$; $N_1 = N_{2\wedge} T$ where $N_2 = \pm \dfrac{X'_\wedge X''}{|X'_\wedge X''|}$

The $\pm$ sign depends on the choice of direction for N_2 and the corresponding sign of K_1. The signs are chosen so that $X'_\wedge X'' = K_1 N_2 (ds/dt)^3$.

3. Find T, N_1, N_2, K_1, and K_2 for the following curves

(a) $x = t$, $y = t^2$, $z = t^3$

(b) $x = a(3t - t^3)$, $y = 3at^2$, $z = a(3t + t^3)$

(c) $x = f(s)$, $y = g(s)$, $z = \cos\alpha$ where α is constant. Show that $K_1/K_2 = \tan\alpha$.

4. Prove: If $K_2 \equiv 0$ for a curve in R^3, then the curve lies in a plane. (By assumption $K_1 \neq 0$ at any point).

5. Show that the Taylor's expansion for a curve in R^3 has the initial expression:

$$F(s_o + h) = F(s_o) + sT + 1/2\, K_1 s^2 N$$

$$+ 1/6\, s^3(-K_1{}^2\, T + K_1'\, N_1 + K_1K_2N_2) + \dots$$

6. If $g_{\alpha\beta}$ are components of the metric tensor in terms of the parameters u and $\bar{g}_{\delta\varepsilon}$ are the components in terms of the v's, and if $g = \det(g_{\alpha\beta})$, $\bar{g} = \det(\bar{g}_{\delta\varepsilon})$, show that $g = \bar{g}\, J^2$ where $J = \det(\partial v^\alpha/\partial u^\beta)$. Hence $\sqrt{g} = \sqrt{\bar{g}}\, |J|$. From this result and the theorem for the change of variables in a multiple integral show that

$$\int\sqrt{g}\, du = \sqrt{\bar{g}}\, dv$$

where the integrals are taken over corresponding domains.

7. A surface of revolution in R^3 has the parametric equations:

$$x = u \cos v,\ y = u \sin v,\ z = f(u)$$

Find $\sqrt{g}$ and show that this gives a formula for the area of the surface equivalent to that proved in elementary calculus.

8. Show that $b_{\alpha\beta} = \bar{b}_{\delta\varepsilon} \dfrac{\partial v^\delta}{\partial u^\alpha} \dfrac{\partial v^\varepsilon}{\partial u^\beta}$. This shows that $b_{\alpha\beta}$ are components of a tensor of type (0, 2).

9. Define $g_{\alpha\beta,\delta} = \dfrac{\partial g_{\alpha\beta}}{\partial u^\delta} - \Gamma^\varepsilon_{\alpha\delta}\, g_{\varepsilon\beta} - \Gamma^\varepsilon_{\beta\delta}\, g_{\alpha\varepsilon}$. Prove that

$$g_{\alpha\beta,\delta} = 0.$$

10. Find the law of transformation under change of parameters for $g^{\alpha\beta}$. Hence $g^{\alpha\beta}$ are components of a tensor of type (2,0).

Exercises 11 to 14 refer to a hypersurface in R^n.

11. Prove that $\nabla_X(Y + Z) = \nabla_X Y + \nabla_X Z$

12. Prove that $\nabla_{X+Y}(Z) = \nabla_X Z + \nabla_Y Z$

13. Prove that $\nabla_{fX}(Y) = f\nabla_X Y$

14. Prove that $\nabla_X(fY) = (\nabla_X f)\, Y + f\nabla_X Y$

In Exercises 15 to 20 find the total curvature K_T at an arbitrary point. These are all surfaces in R^3.

15. Plane: $x = a$, $y = u$, $z = v$.

16. Cylinder: $x = r \cos u$, $y = r \sin u$, $z = v$.

17. Sphere: $x = r \sin u \cos v$, $y = r \sin u \sin v$, $z = r \cos u$.

18. Torus: $x = (a + b \cos u) \cos v$, $y = (a + b \cos u) \sin v$, $z = b \sin u$.

19. Tangent Developable: Let C be defined by $Z = Z(s)$, $T_c = dZ/ds$. Then the surface is defined by

$$X(s,u) = Z(s) + u\, T_c \quad \text{for} \quad u \neq 0.$$

20. Pseudosphere: $x = u \cos v$, $y = u \sin v$,

$$z = a \log \frac{a + \sqrt{a^2 - u^2}}{u} - \sqrt{a^2 - u^2}$$

21. Derive the Gauss and Codazzi equations as the integrability conditions for the Frenet equations of a hypersurface.

Chapter 5

Multilinear Algebra

5.1 Introduction. This chapter is an extension of the ideas of Linear Algebra which covers concepts to be needed later. The treatment is limited to those parts of multilinear algebra that we will have occasion to use. Throughout the chapter we shall be working with a fixed, finite dimensional vector space V over the reals with a basis $(e_1,\dots,e_n)$.

5.2 Dual Vectors.

Definition. A dual vector is a real valued function $L\colon V\to R$ which is linear over the reals. That is

$$L(av_1 + bv_2) = aL(v_1) + bL(v_2)$$

where v_1 and v_2 are vectors in V. Observe that L is defined if we know $L(e_i)$ for each basis element e_i of V.

Theorem 1. The set of dual vectors over V is a vector space, V^*.

Proof. Define $(L_1+L_2)v = L_1(v) + L_2(v)$; $(aL)(v) = a[L(v)]$. The zero vector L_0 is the function such that $L_0(v) = 0$ for all v. With these definitions it is easy to verify that V^* is a vector space over the reals.

Let the dual vectors $f^1,\dots,f^n$ be defined by

$$f^i(e_j) = \delta^i_j$$

In view of this definition we have the result:

Theorem 2. The set $\{f^i\}$ is a basis for V^*. It is called the canonical basis of V^*.

Proof. (1) Consider an arbitrary element, L, of V^* defined by $L(e_i) = a_i$. Then $L(e_i) = a_j\delta^j_i = a_j f^j(e_i)$. Hence $L = a_j f^j$; and so $\{f^i\}$ span V^*.

(2) f^i are independent. If $c_i f^i = 0$, then $c_i f^i(e_j) = 0$, or $c_i\delta^i_j = 0$, or $c_j = 0$.

If we change the basis in V such that $e_j = A^i_j \overline{e}_i$, we ask what effect this has on the canonical basis of V^*. By definition

$$f^i(e_j) = \delta^i_j$$

$$f^i(A^k_j \overline{e}_k) = \delta^i_j$$

$$f^i(\overline{e}_k) = (A^j_k)^{-1}\,\delta^i_j = (A^i_k)^{-1} = (A^i_j)^{-1}\,\delta^j_k = (A^i_j)^{-1}\,\overline{f}^j(\overline{e}_k)$$

Hence

$$f^i = \overline{f}^j(A^i_j)^{-1}$$

This gives the result:

Theorem 3. If $e = A\overline{e}$ is a change of basis in V, then $f = \overline{f}A^{-1}$ is the corresponding change in the canonical basis of V^*.

5.3 Tensors on V.

Definition. A tensor $T(r,s)$ of type (r,s) on V is a mapping:

$$T(r,s):\ \underbrace{V^*\times\cdots\times V^*}_{r\ \text{factors}} \times \underbrace{V\times\cdots\times V}_{s\ \text{factors}} \to R$$

which is linear over the reals in each factor. Thus a tensor is a generalization of a dual vector, which is a tensor of type (0,1).

We wish to show that the set of tensors of fixed type (r,s) is a vector space. To do so we will establish a basis.

Let the tensors $E^{(p_1 \ldots p_s)}_{(q_1 \ldots q_r)}$ be defined by:

$$E^{(p_1 \ldots p_s)}_{(q_1 \ldots q_r)}\left(f^{i_1},\ldots,f^{i_r}, e_{j_1},\ldots e_{j_s}\right) = \delta^{i_1}_{q_1}\ldots\delta^{i_r}_{q_r}\,\delta^{p_1}_{j_1}\ldots\delta^{p_s}_{j_s}$$

Then we have the result:

<u>Theorem 4</u>. The space of all tensors of type $T(r,s)$ is a vector space for which $E^{(p_1 \ldots p_s)}_{(q_1 \ldots q_r)}$ constitute a basis.

<u>Proof</u>. Let $T(r,s)(f^{i_1},\ldots,f^{i_r}, e_{j_1},\ldots,e_{j_s}) = T^{i_1 \ldots i_r}_{j_1 \ldots j_s}$

$$= T^{q_1 \ldots q_r}_{p_1 \ldots p_s}\,\delta^{i_1}_{q_1}\ldots\delta^{i_r}_{q_r}\;\delta^{p_1}_{j_1}\ldots\delta^{p_s}_{j_s}$$

$$= T^{q_1 \ldots q_r}_{p_1 \ldots p_s}\,E^{(p_1 \ldots p_s)}_{(q_1 \ldots q_r)}(f^{i_1},\ldots,f^{i_r}, e_{j_1},\ldots,e_{j_s})$$

So $T(r,s) = T^{q_1 \ldots q_r}_{p_1 \ldots p_s}\,E^{(p_1 \ldots p_s)}_{(q_1 \ldots q_r)}$

Hence $E^{(p_1 \ldots p_s)}_{(q_1 \ldots q_r)}$ span the space of tensors $T(r,s)$.

<u>Remark</u>. To avoid notational confusion, we have used parentheses around those suffixes that indicate the <u>name</u> of the tensor; for example $E^{(p_1 \ldots p_s)}_{(q_1 \ldots q_s)}$ are the names of a set of tensors.

The similar notation without parentheses indicates the <u>components</u> of a tensor. Thus $T^{q_1 \cdots q_r}_{p_1 \cdots p_s}$ are components of the tensor $T(r,s)$.

As in 5.2 we can prove that the E's are independent. Thus the vector space of tensors $T(r,s)$ has dimension n^{r+s}.

The components of $T(r,s)$ with respect to this basis are $T^{q_1 \cdots q_r}_{p_1 \cdots p_s}$.

If we make the change of basis $e_j = A^i_j \bar{e}_i$, $f^i = \bar{f}^j B^i_j$ where $B^i_j = (A^i_j)^{-1}$, we find that

$$T^{i_1 \cdots i_r}_{j_1 \cdots j_s} = T(r,s)\ (f^{i_1}, \ldots, f^{i_r}, e_{j_1}, \ldots, e_{j_s})$$

$$= T(r,s)\ (\bar{f}^{q_1}, \ldots, \bar{f}^{q_r}, \bar{e}_{p_1}, \ldots, \bar{e}_{p_s})\ B^{i_1}_{q_1} \cdots B^{i_r}_{q_r} A^{p_1}_{j_1} \cdots A^{p_s}_{j_s}$$

$$= \bar{T}^{q_1 \cdots q_r}_{p_1 \cdots p_s} B^{i_1}_{q_r} \cdots B^{i_r}_{q_r} A^{p_1}_{j_1} \cdots A^{p_s}_{j_s}$$

So we have Theorem 5:

<u>Theorem 5</u>. Under a change of basis $e_j = A^i_j \bar{e}_i$ in V, the components of tensors of type (r,s) transform by the following formula, where $B^i_j = (A^j_i)^{-1}$:

$$T^{i_1 \cdots i_r}_{j_1 \cdots j_s} = \bar{T}^{q_1 \cdots q_r}_{p_1 \cdots p_s} B^{i_1}_{q_1} \cdots B^{i_r}_{q_r} A^{p_1}_{j_1} \cdots A^{p_s}_{j_s}$$

In the terminology of tensor analysis the tensor $T(r,s)$ with components $T^{i_1 \cdots i_r}_{j_1 \cdots j_s}$ is <u>contravariant</u> of order r and <u>covariant</u> of order s. This terminology appears to be just the opposite of what logic would dictate, but it is too well established to alter here.

5.4 <u>The Double Dual Vector Space V^{**}</u>. Since V^* is a vector space, we can consider its dual, V^{**}, which is the vector space of real valued linear functions defined on V^*. If v is a vector in V, and w is a vector in V^*, then $w(v)$ is a real number. We can define a vector x in V^{**} by requiring that $x(w) = w(v)$ for all w in V^*. Thus we have a single valued mapping ϕ: $v \to x$ or $x = \phi(v)$ so that $w(v) = [\phi(v)]w$.

(1) <u>ϕ is linear</u>. For:

$$
\begin{aligned}
[\phi(av_1 + bv_2)](w) &= w(av_1 + bv_2) \\
&= w(v_1) + b\, w(v_2) \\
&= a\, \phi(v_1)(w) + b\, \phi(v_2)(w)
\end{aligned}
$$

(2) <u>ϕ is one-to-one</u>. For if $\phi(v_1) = \phi(v_2)$, then $w(v_1)$ $= w(v_2)$ or $w(v_1 - v_2) = 0$ for all w. Hence $v_1 = v_2$.

Thus there is a natural isomorphism between V and V^{**}, and we are justified in identifying these two spaces. In particular $\phi(e_i) = e_i$ is a basis for V^{**} and so $e_i(f^j) = f^j(e_i) = \delta^i_j$. Thus we may consider a vector of V, say $v = v^i e_i$ to be a tensor of type $(1,0)$.

5.5 <u>Products of Tensors</u>. We define a product which maps a pair of tensors: $T(r,s)$ and $T(p,q)$ into $T(r+p,\ s+q)$. This is known as the <u>tensor product</u>. The definition is a natural one as follows:

<u>Definition</u>. If $T(r,s)$ has components relative to the standard basis $T^{i_1 \cdots i_r}_{j_1 \cdots j_s}$ and $T(p,q)$ has components $T^{u_1 \cdots u_p}_{v_1 \cdots v_q}$, then the tensor product $T(r,s) \otimes T(p,q)$ is the tensor whose value on:

$$(f^{i_1}, \ldots, f^{i_r},\ f^{u_1}, \ldots, f^{u_p},\ e_{j_1}, \ldots, e_{j_s},\ e_{v_1}, \ldots, e_{v_q}) \quad \text{is}$$

$$T^{i_1 \cdots i_r}_{j_1 \cdots j_s}\ T^{u_1 \cdots u_p}_{v_1 \cdots v_q}$$

where this last product is a product of real numbers.

<u>Examples</u>. 1. $e_{q_1} \otimes e_{q_2}$ is the tensor such that

$$(e_{q_1} \otimes e_{q_2})(f^{i_1}, f^{i_2}) = \delta^{i_1}_{q_1}\ \delta^{i_2}_{q_2}$$

2. $f^{p_1} \otimes f^{p_2}$ is the tensor such that

$$(f^{p_1} \otimes f^{p_2})(e_{j_1}, e_{j_2}) = \delta^{p_1}_{j_1}\ \delta^{p_2}_{j_2}$$

3. $e_{q_1} \otimes f^{p_1}$ is the tensor such that

$$(e_{q_1} \otimes f^{p_1})(f^{i_1}, e_{j_1}) = \delta^{i_1}_{q_1}\ \delta^{p_1}_{j_1}$$

4. If v and w are vectors in V and

$$v = v^r e_r, \quad w = w^s e_s, \quad \text{then}$$

$$v \otimes w = v^r w^s (e_r \otimes e_s)$$

In R^3 this corresponds to the matrix product

$$\begin{pmatrix} v^1 \\ v^2 \\ v^3 \end{pmatrix} (w^1 \ w^2 \ w^3) = \begin{pmatrix} v_1w_1 & v_1w_2 & v_1w_3 \\ v_2w_1 & v_2w_2 & v_2w_3 \\ v_3w_1 & v_3w_2 & v_3w_3 \end{pmatrix}$$

In the older literature (still extant in a few books on Vector Analysis), this is known as the dyadic product of two vectors.

Using this definition we can prove Theorem 6.

Theorem 6. Any tensor of type (r,s) is the sum (with real coefficients) of tensor products of r tensors of type $(1,0)$, namely vectors, and s tensors of type $(0,1)$, namely dual vectors.

Proof. First we show that

$$E^{(p_1 \cdots p_s)}_{(q_1 \cdots q_r)} = e_{q_1} \otimes \cdots \otimes e_{q_r} \otimes f^{p_1} \otimes \cdots \otimes f^{p_s}$$

where $E^{(p_1 \cdots p_s)}_{(q_1 \cdots q_r)}$ are the basis of the space of tensors of type (r,s) defined in Section 5.3. This follows since

$$(e_{q_1} \otimes \cdots \otimes e_{q_r} \otimes f^{p_1} \otimes \cdots \otimes f^{p_s})(f^{i_1}, \ldots, f^{i_r}, e_{j_1}, \ldots, e_{j_s})$$

$$= \delta^{i_1}_{q_1} \cdots \delta^{i_r}_{q_r} \delta^{p_1}_{j_1} \cdots \delta^{p_s}_{j_s}$$

$$= E^{(p_1 \cdots p_s)}_{(q_1 \cdots q_s)} (f^{i_1}, \ldots, f^{i_r}, e_{j_1}, \ldots, e_{j_s})$$

and $e_{q_1}, \ldots, e_{q_r}$ are tensors of type (1,0) and $f^{p_1}, \ldots, f^{p_s}$ are tensors of type (0,1).

Finally any tensor of type $T(r,s)$ is a linear combination (with real coefficients) of $E^{(p_1 \ldots p_s)}_{(q_1 \ldots q_r)}$.

Because of this theorem, the vector space of tensors of type (r,s) is called the tensor product of the vector spaces:

$$\underbrace{V \otimes \cdots \otimes V}_{r \text{ factors}} \otimes \underbrace{V^* \otimes \cdots \otimes V^*}_{s \text{ factors}}$$

5.6 <u>The Contraction Operator</u>. This is an operation which maps a tensor of type (r,s) into a tensor of type $(r-1,s-1)$. It is most easily defined in terms of components. For example if T^i_j are the components of a tensor of type (1,1), then its contraction is the tensor of type (0,0) with the single component $\sum_i T^i_i$. In general if we have the components of a tensor of type (r,s): $T^{i_1 \ldots i_r}_{j_1 \ldots j_s}$ we may equate any upper index with any lower index and sum on this index. For example two possible contractions are:

$$\sum_i T^{ii_2 \ldots i_r}_{ij_2 \ldots j_s} \quad \text{and} \quad \sum_i T^{ii_2 \ldots i_r}_{j_1 \ldots j_{s-1} i}$$

To show that this operation applies to the tensors themselves and not to just the components, we must show how it is affected by a change of basis. To keep the formulas as simple as possible, consider

$$T^{ij}_k \text{ and its contraction } \sum_i T^{ij}_i$$

Then

$$T_k^{ij} = \overline{T}_t^{rs} B_r^i B_s^j A_k^t$$

$$\sum_i T_i^{ij} = \sum_i \left(\overline{T}_t^{rs} B_r^i B_s^j A_i^t\right)$$

$$= \sum_t \overline{T}_t^{rs} B_s^j \delta_r^t$$

$$= \sum_t \overline{T}_t^{ts} B_s^j$$

which is the law of transformation of the components of a tensor of type (1,0).

The operations of tensor product and contraction may be combined as in the example (in components):

$$T_k^{ij} V^k$$

which is the product $T_k^{ij} V^r$ contracted on k and r.

Thus $T_k^{ij}V^k$ are the components of a tensor of type (2,0).

5.7 The Exterior Algebra of Covectors.

Definition. A p-covector is a tensor $T(0,p)$ of type $(0,p)$ such that

$$T(0,p)(v_1,\ldots,v_i,\ldots,v_j,\ldots,v_p)$$

$$= -T(0,p)(v_1,\ldots,v_j,\ldots,v_i,\ldots,v_p)$$

for each pair of vectors v_i and v_j. This means that the components of $T(0,p)$ are skew-symmetric in each pair of indices. Such a tensor is often called alternating. For example:

$$T(0,2)(e_i,e_j) = -T(0,2)(e_j,e_i); \quad \text{so} \quad T_{ij} = -T_{ji}$$

Similarly for a 3-covector

$$T_{ijk} = -T_{jik} = -T_{kji} = -T_{ikj}$$

We shall use the notation w^p to indicate a p-covector. By convention we say that a real number (a scalar) is a 0-covector and that a dual vector is a 1-covector.

Next we consider the so-called "Exterior Algebra" of covectors. We begin with the exterior product of two 1-covectors w_1 and w_2.

<u>Definition</u>. The exterior product $w_1 \wedge w_2$ of w_1 and w_2 is the skew-symmetric part of $w_1 \otimes w_2$. That is,

$w_1 \wedge w_2 = w_1 \otimes w_2 - w_2 \otimes w_1$. Thus $w_1 \wedge w_2 = -w_2 \wedge w_1$ and $w_1 \wedge w_1 = 0$. Thus $w_1 \wedge w_2$ is a 2-covector. In this definition we have omitted the factor 1/2 which frequently appears. Similarly in later formulas we omit factors such as $1/p!$ Because of this convention, our formulas differ by various constants from those that appear in some other treatments of this subject.

The value of $w_1 \wedge w_2$ on a pair of vectors v_1, v_2 is given by:

$$(w_1 \wedge w_2)(v_1,v_2) = w_1(v_1)\, w_2(v_2) - w_1(v_2)\, w_2(v_1)$$

<u>Generalization</u>. If $w_1,\ldots,w_p$ are 1-covectors, we define $w_1\wedge\ldots\wedge w_p$ as the skew-symmetric part of $w_1\otimes\ldots\otimes w_p$. That is:

$$(w_1\wedge\ldots\wedge w_p)(v_1,\ldots,v_p) = [w_1(v_{i_1}),\ldots,w_p(v_{i_p})]\ \delta^{i_1\ldots i_p}_{1\ldots p}$$

where $\delta^{i_1\cdots i_p}_{1\cdots p} = \begin{cases} 1 \text{ if } i_1\cdots i_p \text{ is an even permutation of } 1\cdots p \\ -1 \text{ if } i_1\cdots i_p \text{ is an odd permutation of } 1\cdots p \\ 0 \text{ otherwise.} \end{cases}$

Thus in particular:

$$(f^{i_1}\wedge\cdots\wedge f^{i_p})(e_{j_1},\ldots,e_{j_p}) = \delta^{i_1}_{k_1}\cdots\delta^{i_p}_{k_p}\,\delta^{k_1\cdots k_p}_{j_1\cdots j_p} = \delta^{i_1\cdots i_p}_{j_1\cdots j_p}$$

where $\delta^{i_1\cdots i_p}_{j_1\cdots j_p} = \begin{cases} 1 \text{ if } i_1\cdots i_p \text{ is an even permutation of } j_1\cdots j_p \\ -1 \text{ if } i_1\cdots i_p \text{ is an odd permutation of } j_1\cdots j_p \\ 0 \text{ otherwise.} \end{cases}$

Then we have the result:

Theorem 7. $f^{i_1}\wedge\cdots\wedge f^{i_p}$ $(i_1 < i_2 < \cdots < i_p)$ form a basis for the vector space of p-covectors.

Proof. (1) $f^{i_1}\wedge\cdots\wedge f^{i_p}(i_1 < i_2 < \cdots < i_p)$ are independent. For if

$$\Sigma' C_{i_1\cdots i_p} f^{i_1}\wedge\cdots\wedge f^{i_p} = 0$$

(where Σ' is the sum over $i_1 < i_2 < \cdots < i_p$)

then

$$0 = \Sigma' C_{i_1\cdots i_p} f^{i_1}\wedge\cdots\wedge f^{i_p}(e_{j_1},\ldots,e_{j_p})$$

and

$$0 = \Sigma' C_{i_1\cdots i_p}\,\delta^{i_1\cdots i_p}_{j_1\cdots j_p}$$

$$= C_{j_1\cdots j_p}$$

(2) $f^{i_1}\wedge\cdots\wedge f^{i_p}(i_1 < i_2 < \cdots < i_p)$ span the space of p-covectors.

Consider the p-covector w^p where

$$w^p(e_{i_1},\ldots,e_{i_p}) = T_{i_1\ldots i_p} = \frac{1}{p!} T_{j_1\ldots j_p} \delta^{j_1\ldots j_p}_{i_1\ldots i_p}$$

$$= \frac{1}{p!} T_{j_1\ldots j_p} f^{j_1}\wedge\ldots\wedge f^{j_p}(e_{i_1},\ldots,e_{i_p})$$

So

$$w^p = \frac{1}{p!} T_{j_1\ldots j_p} f^{j_1}\wedge\ldots\wedge f^{j_p} = \Sigma' T_{j_1\ldots j_p} f^{j_1}\wedge\ldots\wedge f^{j_p}$$

The dimension of the space of p-covectors is thus $\binom{n}{p}$.

5.8 <u>Exterior Products of Covectors</u>. Let

Let $w^p = \frac{1}{p!} T_{j_1\ldots j_p} f^{j_1}\wedge\ldots\wedge f^{j_p}$ be a p-covector and

$w^q = \frac{1}{q!} S_{i_1\ldots i_q} f^{i_1}\wedge\ldots\wedge f^{i_q}$ be a q-covector

Then we have the definition:

<u>Definition</u>. $w^p_p \wedge w^q$ is the alternating part of $w^p \otimes w^q$.

So $w^p \wedge w^q$ is a $(p+q)$-covector.

Then

$$w^p \wedge w^q = \frac{1}{p!}\frac{1}{q!} T_{j_1\ldots j_p} S_{i_1\ldots i_q} f^{j_1}\wedge\ldots\wedge f^{j_p} \wedge f^{i_1}\wedge\ldots\wedge f^{i_q}$$

$$= \frac{1}{(p+q)!} U_{k_1\ldots k_{p+q}} f^{k_1}\wedge\ldots\wedge f^{k_{p+q}}$$

where $U_{k_1\ldots k_{p+q}} = T_{j_1\ldots j_p} S_{i_1\ldots i_q} \delta^{j_1\ldots j_p i_1\ldots i_q}_{k_1\ldots k_{p+q}}$

Then we have the theorem:

Theorem 8. $w^p \wedge w^q = (-1)^{pq} w^q \wedge w^p$

$$(w^p \wedge w^q) \wedge w^r = w^p \wedge (w^q \wedge w^r)$$

Thus the set of p-covectors for $p = 1 \ldots n$ is a graded ring. Sums of covectors of the same grade are defined, and products $w^p \wedge w^q$ are of grade $p + q$ and are thus included in the complete set.

5.9 Orientation of V. Our purpose here is to discuss the notion of the orientation of an ordered set of independent vectors $(v_1,\ldots,v_n)$ in V.

Definition. Let $(v_1,\ldots,v_n)$ be an ordered set of independent vectors in V. Then an orientation of $(v_1,\ldots,v_n)$ is a number $\mu = \pm 1$ such that

$$\mu(v_1,\ldots,v_i,\ldots,v_j,\ldots,v_n) = -(v_1,\ldots,v_j,\ldots,v_i,\ldots,v_n)$$

for each pair of vectors v_i and v_j.

To do this in a systematic fashion, we first choose an ordered basis for $v_i(e_1,\ldots,e_n)$ which we call the fundamental basis and define $\mu(e_1,\ldots,e_n) = +1$. For example in R^n we choose this fundamental basis to be the usual basis of Section 1.1.

To determine the orientation of an arbitrary ordered set of vectors $(v_1,\ldots,v_n)$ we express each of these in terms of the fundamental basis: $v_i = v_i^j e_j$. Then by definition:

Definition. If $(v_1,\ldots,v_n)$ is an ordered set of independent vectors in V with components v_i^j in terms of the fundamental basis, then

$$\mu(v_1, \ldots, v_n) = \text{sign det } (v_i^j).$$

Of course, we may wish to express $\mu(v_1, \ldots, v_n)$ in terms of their components relative to an arbitrary ordered basis $(\bar{e}_1, \ldots, \bar{e}_n)$, such that $e_i = A_i^j \bar{e}_j$. Let the components of v_i with respect to the $\bar{e}$ basis be $\bar{v}_i^j$. Then it follows easily that

$$\mu(v_1, \ldots, v_n) = \text{sign det } (\bar{v}_i^j) \times \text{sign det } (A_i^j)$$

Even so, we remind you that $\mu(v^1, \ldots, v^n)$ depends not only on $(v^1, \ldots, v^n)$ but on the choice of the fundamental basis.

An orientation of V induces an orientation on the dual space V^*. For we choose the fundamental basis of V^* to be the ordered canonical basis $(f^1, \ldots, f^n)$ derived from the ordered basis $(e_1, \ldots, e_n)$ of V.

Exercises

1. Show that for vectors in V:

$$v \otimes (w_1 + w_2) = v \otimes w_1 + v \otimes w_2$$

$$(v_1 + v_2) \otimes w = (v_1 \otimes w) + (v_2 \otimes w)$$

$$\lambda(v \otimes w) = (\lambda v) \otimes w = v \otimes (\lambda w)$$

2. Evaluate $w^p \wedge w^p$ where w^p is a p-covector when p is odd. What is true when p is even?

3. Evaluate $w^p \wedge w^q$ when $p + q > n$.

4. Show that $(w^p + w^q) \wedge w^r = (w^p \wedge w^r) + (w^q \wedge w^r)$

5. If w^p is a p-covector and v is a vector, we define the "interior product" $i(v)w^p$ as the (p-1)-covector whose value on the set of p-1 vectors $v_1,\ldots,v_{p-1}$ is

$$[i(v)w^p]\,[v_1,\ldots,v_{p-1}] = w^p[v,v_1,\ldots,v_{p-1}]$$

Show that

(a) $i(v)df = df(v) = vf$ for a function f

(b) $i(v)(w^p \wedge w^q) = [i(v)w^p] \wedge w^q + (-1)^p w^p \wedge [i(v)w^q]$

6. Prove that the 1-covectors $w_1,\ldots,w_p$ are linearly dependent iff

$$w_1 \wedge\ldots\wedge w_p = 0$$

7. Prove: If the 1-covectors $w_1,\ldots,w_p$ are linearly independent and if $\pi_1\ldots\pi_r$ is another set of 1-covectors such that

7. (continued)

$$\sum_{s=1}^{p} \pi_s \wedge w_s = 0$$

then
$$\pi_s = \sum_{t=1}^{r} a_{st} w_t$$

where $a_{st} = a_{ts}$.

8. Let V and W be arbitrary finite dimensional vector spaces with bases v_i and w_α, and corresponding dual bases r^i and s^α. By definition the tensor product $V \otimes W$ is the vector space of bilinear, real-valued functions defined on $V^* \times W^*$. Prove that a basis for $V \otimes W$ is the set: $v_i \otimes w_\alpha$ where

$$(v_i \otimes w_\alpha)(r^j, s^\beta) = \delta_i^j \, \delta_\alpha^\beta$$

9. Prove (see exercise 8) that $V \otimes W$ and $W \otimes V$ are naturally isomorphic.

10. Prove (see exercise 8) that $(U \otimes V) \otimes W$ and $U \otimes (V \otimes W)$ are naturally isomorphic.

11. Define $U \otimes V \otimes W$ from scratch and show that it is naturally isomorphic to $(U \otimes V) \otimes W$ and to $U \otimes (V \otimes W)$.

12. In our definition of the tensor product in 5.5, the definition is given relative to a particular basis for V. Show that the definition is invariant under change of basis.

Exercises 13 to 18 concern determinants. In these

$$\varepsilon_{i_1 \ldots i_n} = \delta^{1 \ldots\ldots n}_{i_1 \ldots i_n}$$

13. Let a^i_j be an $n \times n$ matrix and $a = \det (a^i_j)$. Show that

$$a = \varepsilon_{i_1 \ldots i_n} \, a^{i_1}_{1} \ldots a^{i_n}_{n}$$

using the definition of a determinant.

14. Show that

$$a \, \varepsilon_{i_1 \ldots i_n} = a^{j_1}_{i_1} \ldots a^{j_n}_{i_n} \, \varepsilon_{j_1 \ldots j_n}$$

15. Prove that $(\det A)(\det B) = \det (AB)$ following the outline:

$$a \, b = a \, \varepsilon_{i_1 \ldots i_n} \, b^{i_1}_{1} \ldots b^{i_n}_{n}$$

$$= a^{j_1}_{i_1} \ldots a^{j_n}_{i_n} \, \varepsilon_{j_1 \ldots j_n} \, b^{i_1}_{1} \ldots b^{i_n}_{n}$$

$$= \varepsilon_{j_1 \ldots j_n} (a^{j_1}_{i_1} \, b^{i_1}_{1}) \ldots (a^{j_n}_{i_n} \, b^{i_n}_{1})$$

16. Define the cofactor of a^j_p to be

$$A^p_j = \varepsilon_{i_1 \ldots i_n} \, a^{i_1}_{1} \ldots a^{i_{p-1}}_{p-1} \, \delta^{i_p}_{j} \, a^{i_{p+1}}_{p+1} \ldots a^{i_n}_{n}$$

Then prove that $a^j_q \, A^p_j = a \, \delta^p_q$

17. Suppose that a^i_j are differentiable functions of x^j.

Then $\frac{\partial a}{\partial x^j} = \varepsilon_{i_1 \ldots i_n} \frac{\partial a^{i_1}_1}{\partial x^j} a^{i_2}_2 \ldots a^{i_n}_n + \ldots$

Show that

$$\frac{\partial a}{\partial x^j} = \frac{\partial a^b_c}{\partial x^j} A^c_b$$

18. Let (g_{ij}) be as in Chapter 4 and $g = \det (g_{ij})$.

Show that

$$\frac{\partial g}{\partial x^k} = \frac{\partial g_{ij}}{\partial x^k} g^{ij} g$$

Hence prove that

$$\Gamma^i_{ik} = \frac{1}{2} \frac{\partial \log g}{\partial x^k}$$

19. If P^i are the components of a vector, we define $P^i_{,j}$ to be $\frac{\partial P^i}{\partial x^j} + \Gamma^i_{jk} P^k$. Prove that the contraction

$P^i_{,i} = \frac{1}{\sqrt{g}} \frac{\partial}{\partial x^i} (\sqrt{g}\, P^i)$. This expression is called the divergence of P^i, div P.

20. If F is a function, we define

$$F_{,i,j} = \frac{\partial^2 F}{\partial x^i \partial x^j} - \Gamma^k_{ij} \frac{\partial F}{\partial x^k}$$

The Laplacian of F, $\nabla^2 F$, is defined to be $g^{ij} F_{,i,j}$.

(a) If $P^i = g^{ij} \frac{\partial F}{\partial x^j}$, prove that $\nabla^2 F = P^i_{\ ,i}$;

or that $\nabla^2 F = \text{div grad } F$.

(b) Hence from exercise 19, prove that

$$\nabla^2 F = \frac{1}{\sqrt{g}} \frac{\partial}{\partial x^i} \left(\sqrt{g}\, g^{ij} \frac{\partial F}{\partial x^j} \right)$$

(c) In polar coordinates $ds^2 = dr^2 + r^2 d\theta^2$ so $g_{rr} = 1$, $g_{r\theta} = 0$, $g_{\theta\theta} = r^2$. Find $\nabla^2 F(r,\theta)$ in polar coordinates.

The expression in (b) for $\nabla^2 F$ is of great importance in applied mathematics, for it saves an enormous amount of calculation.

Chapter 6

Differentiable Manifolds

6.1 Introduction. An abstract differentiable manifold is a generalization of an imbedded submanifold of R^n. In this case, however, there is no R^n in the picture, so we must start from scratch.

Definition. An n-dimensional C^p $(p\geq 1)$ manifold, M, is the following object:

(a) M is a topological space (usually assumed to be Hausdorff and paracompact) with a covering of open sets $\{U_\alpha\}$.

(b) Each U_α is homeomorphic to the interior of a ball B_α in R^n. Write the homeomorphism:

$$\phi^\alpha: U_\alpha \to B_\alpha$$

The pair (U_α, ϕ_α) is called a coordinate system or a chart. The totality of charts is an atlas.

(c) Suppose that $U_\alpha \cap U_\beta \neq \emptyset$. Let $\overline{\phi}_\alpha$, $\overline{B}_\alpha$, $\overline{\phi}_\beta$, and $\overline{B}_\beta$ be the restrictions of ϕ_α, B_α, ϕ_β, and B_β such that

$$\overline{\phi}_\alpha: U_\alpha \cap U_\beta = \overline{B}_\alpha \quad \text{and} \quad \overline{\phi}_\beta: U_\alpha \cap U_\beta = \overline{B}_\beta$$

are homeomorphisms. Then there is a map:

$$\psi_{\beta\alpha}: \overline{B}_\alpha \to \overline{B}_\beta$$

defined by:

$$\psi_{\beta\alpha} = \overline{\phi}_\beta \circ \overline{\phi}_\alpha^{-1}.$$

We require that each $\psi_{\beta\alpha}$ be a C^p function with non-vanishing Jacobian. They are automatically one-to-one.

6.2 Functions on M. Let f be a function $f: M \to R^1$. There is not enough structure on M itself to discuss the differentiability of f, so we introduce the function $F_\alpha: B_\alpha \to R^1$ defined by:

$$F_\alpha(x) = (f \circ \phi_\alpha^{-1})(x) \qquad \text{where} \quad x \in B_\alpha$$

From this it follows that

$$f(P) = (F_\alpha \circ \phi_\alpha)(P) \qquad \text{where} \quad P \in U_\alpha$$

Thus there is a one-to-one correspondence between functions $F_\alpha(x)$ defined in B_α and functions $f|U_\alpha$ defined in U_α.

To define the differentiability of f, we say:

Definition. $f|U_\alpha$ is a C^p function on U_α iff F_α is C^p on B_α.

More generally we have:

Definition. Let M be a C^p manifold. Then f is a $C^r (r \leq p)$ function on M iff f is C^r on each open set of the covering $\{U_\alpha\}$ of M.

This definition is consistent, for:

$$F_\alpha = f \circ \phi_\alpha^{-1} = f \circ (\phi_\beta^{-1} \circ \psi_{\beta\alpha}) = (f \circ \phi_\beta^{-1}) \circ \psi_{\beta\alpha}$$

$$= F_\beta \circ \psi_{\beta\alpha}$$

Since F_β is of class C^r and since $\psi_{\beta\alpha}$ is of class C^p $(p \geq r)$, it follows that F_α is of class C^r. This proves the consistency of our definition.

Note that functions of class C^r cannot be defined on M if M is of class C^p with $p < r$.

6.3 Tangent Vectors on M. Since M is not assumed to be a subspace of R^n, the notion of a (tangent) vector to M at a point P must be defined in a new way. There are several ways to do this. We adopt a definition based on directional derivatives. See exercise 12, Chapter 1.

First let M be of class C^∞ or C^ω.

Definition. A vector X_P at a point $P \in M$ is a mapping from the set of C^∞ or C^ω functions defined on M near P into R^1: $X_P: f \to R^1$ having the two properties:

(1) $\quad X_P(af + bg) = aX_Pf + bX_Pg$

where a and b are constants. Thus X_P is "linear over the constants."

(2) $\quad X_P(fg) = f(P) \cdot X_Pg + g(P) \cdot X_Pf$

Thus X_P is a differentiation.

The set of all X_P at P is a vector space if we choose the zero vector to be $O_P: f \to 0$ for all f.

Definition. The vector space of vectors at $P \in M$ is called the tangent space $T_P(M)$ to M at P.

In order to find the dimension of $T_P(M)$ and a basis for it, we need to map everything into B_α by ϕ_α and work in $B_\alpha \subset R^n$. Let $P \in U_\alpha$ and $\phi_\alpha(P) = 0 \in B_\alpha$. We already have defined $F_\alpha = f \circ \phi_\alpha^{-1}$ as a function defined on B_α near 0.

Now at $0 \in B_\alpha$ we can use the above definition to define a vector Y_0 in R^n corresponding to each X_P on M.

<u>Definition</u>. $Y_0 F_\alpha = X_P f$. Thus Y_0 maps functions F_α defined in B_α near 0 into the reals.

There is thus a one-to-one linear correspondence

$$X_P \to Y_0$$

which sends $0_P \to 0_0$, and thus preserves linear independence. We now need three lemmas:

<u>Lemma 1</u>. Let $F = f \circ \phi_\alpha^{-1}$ and $G = g \circ \phi_\alpha^{-1}$ be defined in B_α. Then

(1) $Y_0(aF + bG) = aY_0 F + bY_0 G$

(2) $Y_0(FG) = F(0) \cdot Y_0 G + G(0) \cdot Y_0 F$

This verifies that Y_0 is a vector.

<u>Proof</u>.

(1) $aF + bG = (af + bg) \circ \phi_\alpha^{-1}$

So

$$Y_0(aF + bG) = X_P(af + bg) = aX_P f + bX_P g = aY_0 F + bY_0 g$$

(2) $FG = (fg) \circ \phi_\alpha^{-1}$

So

$$Y_0(FG) = X_P(fg) = f(P) \cdot X_P g + g(P) \cdot X_P f$$

$$= F(0) \cdot Y_0 G + G(0) \cdot Y_0 F$$

<u>Lemma 2</u>. $Y_0(a) = 0$ for <u>a</u> constant.

From (2) $Y_0(1 \cdot 1) = 1 \cdot Y_0(1) + 1 \cdot Y_0(1)$. So $Y_0(1) = 0$

From (1) $Y_o(a) = Y_o(a \cdot 1) = a\, Y_o(1) = 0$

<u>Lemma 3</u>. $Y_o(F) = \left(\frac{\partial F}{\partial x^i}\right)_o \lambda^i$ where $\lambda^i = Y_o(x^i)$

By Theorem 10, Chapter 1

$$F(x) = F(0) + x^i g_i(x)$$

where $g_i(x)$ are C^∞ or C^ω and $g_i(0) = (\partial F/\partial x^i)_o$.

Hence $Y_o(F) = Y_o[F(0)] + Y_o[x^i g_i(x)]$

$$= 0 + x^i(0)\, Y_o[g_i(x)] + g_i(0)\, Y_o(x^i)$$

$$= 0 + 0 + \left(\frac{\partial F}{\partial x^i}\right)_o \lambda^i$$

Thus we have the result:

<u>Theorem 1</u>. The n vectors

$$\left(\frac{\partial}{\partial x^1}\right)_o \cdots \left(\frac{\partial}{\partial x^n}\right)_o$$

span $T_o(B_\alpha)$.

Moreover, these vectors are linearly independent. For if

$$\left(c^i \frac{\partial}{\partial x^i}\right)_o F = 0 \quad \text{for all } F,$$

then if $F = x^1$ it follows that $c^1 = 0$. Similarly all $c^i = 0$. So we have Theorem 2.

Theorem 2. The vector space $T_o(B_\alpha)$ is n-dimensional and $\{(\partial/\partial x^i)_o\}$ is a basis.

Because of the correspondence $Y_o F_\alpha = X_P f$ we have the result:

Theorem 3. The vector space $T_P(M)$ is n-dimensional and

$(X_i)_P f = [\frac{\partial}{\partial x^i}(f \circ \phi_\alpha^{-1})]_{x=0}$ is a basis.

6.4 Tangent Vectors to a C^p Manifold. When M is of class C^p $(1 \leq p < \infty)$ the definition above of a tangent vector still applies and the set of such vectors is again a vector space. Unfortunately it is infinite dimensional and thus clearly does not give us what we want. [See A. G. Walker and W. F. Newns "Tangent Planes to a Differentiable Manifold", Journal of the London Mathematical Society, vol. 31 (1956) pp. 401-7.]

In order to remedy this, we introduce a more complicated definition of a tangent vector on such manifolds. Let $G(y^1,\ldots,y^n)$ be a C^p function $R^n \to R^1$, and f^i for $i = 1\ldots n$ be C^p functions $M \to R^1$ defined near P. Then $G(f^1,\ldots,f^n)$ is a C^p function $M \to R^1$ defined near P.

Definition. A tangent vector at P is a mapping $X_P f$ of the C^p functions defined near P into R^1 satisfying:

$$(3) \quad X_P[G(f^1,\ldots,f^n)] = \left(\frac{\partial G}{\partial y^i}\right)_{f(P)} \cdot X_P(f^i)$$

Remark. Observe that properties (1) and (2) of our earlier definition are special cases of (3).

We wish to prove the result:

Theorem 4. The vector space $T_P(M)$ is n-dimensional.

Proof. As before we define Y_o at $O \in B_\alpha$ by $Y_o F_\alpha = X_P f$. Let $F^i = f^i \circ \phi_\alpha^{-1}$. Then

$$G(F^1,\ldots,F^n) = G(f^1,\ldots,f^n) \circ \phi_\alpha^{-1}$$

and

$$Y_o[G(F^1,\ldots,F^n)] = X_P[G(f^1,\ldots,f^n)]$$

$$= \left(\frac{\partial G}{\partial y^i}\right)_{f(P)} \cdot X_P f^i$$

$$= \left(\frac{\partial G}{\partial y^i}\right)_{F(O)} \cdot Y_o F^i$$

In particular let $F^1 = x^1,\ldots,F^n = x^n$. Then

$$Y_o G(x^1,\ldots,x^n) = \left(\frac{\partial G}{\partial x^i}\right)_O \cdot Y_o(x^i)$$

$$= \left(\frac{\partial G}{\partial x^i}\right)_O \lambda^i \quad \text{where } \lambda^i = Y_o(x^i).$$

Since $G(x)$ is an arbitrary C^p function on B_α, it follows that

$$Y_o(F^i) = \left(\frac{\partial F^i}{\partial x^j}\right)_O \lambda^j$$

and $Y_o[G(F^1,\ldots,F^n)] = \left(\frac{\partial G}{\partial y^i}\right)_{F(O)} \left(\frac{\partial F^i}{\partial x^j}\right)_O \lambda^j = \frac{\partial}{\partial x^i}[G(F^1,\ldots,F^n)]_O \lambda^i$

Thus $\left\{\frac{\partial}{\partial x^i}\right\}_O$ is a basis of $T_O(B_\alpha)$, and this space is n-dimensional. Hence the space $T_P(M)$ is also n-dimensional with the same basis as in Theorem 3.

6.5 The Tangent Bundle T(M).

Definition. The tangent bundle $T(M)$ is the set $\{T_P(M) | P \in M\}$. It is thus the collection of all the tangent spaces at all points of M.

In R^n, this results in the following situation. As a vector space, R^n has a distinguished point, the origin. Moreover, $T_o(R^n) = R^n$. But $T_P(R^n)$ is also R^n where the origin of R^n is now at P. Since $T_P(R^n)$ may be regarded as a translate of $T_o(R^n)$, a basis in $T_o(R^n)$ induces a parallel basis in $T_P(R^n)$.

Vectors "at a point $P \in R^n$" are thus elements of $T_P(R^n)$ and vectors with the same components at $T_o(R^n)$ and $T_P(R^n)$ are parallel.

In some treatments parallel vectors at different points are called "equivalent" and a vector is defined as the corresponding equivalence class. This use is common in Physics.

On a general manifold this concept of parallelism does not exist, and there is apparently no way to identify vectors defined at different points. A way to do this, when possible, is one of the major concepts of differential geometry (Section 6.16).

We can define a topological structure on $T(M)$ in a fashion such that $T(M)$ is a manifold. Let U_α be a coordinate neighborhood of M with coordinates $x^1,\ldots,x^n$. Then at any point, P, of U_α a basis for $T_P(M)$ is $\left(\frac{\partial}{\partial x^1},\ldots,\frac{\partial}{\partial x^n}\right)$, and an element of $T_P(M)$ can be written $\lambda^i(x)\left(\frac{\partial}{\partial x^i}\right)_P$. Note that we are now simplifying our language by identifying vectors on U_α with the corresponding vectors on B_α.

Thus the bundle over U_α is the Cartesian product: $U_\alpha \times R^n$ whose ordered pairs are (x^i, λ^i). This is clearly homeomorphic to an open set in $R^n \times R^n$. So if we take the open sets of $T(M)$ to be $U_\alpha \times R^n$, it follows that $T(M)$ is a manifold.

As defined, $T(M)$ is a local product, but it is not a product in the large. For if P is an element of $U_\alpha \cap U_\beta$ where x^i are coordinates of U_α and $\bar{x}^i$ are coordinates of U_β, two bases for $T_P(M)$ are defined:

$$\{(\partial/\partial x^i)_P\} \quad \text{and} \quad \{(\partial/\partial \bar{x}^i)_P\}$$

Since

$$\frac{\partial}{\partial x^i} = \frac{\partial \bar{x}^j}{\partial x^i} \frac{\partial}{\partial \bar{x}^j}$$

it follows that

$$(1) \qquad \bar{\lambda}^j = \lambda^i \frac{\partial \bar{x}^j}{\partial x^i}$$

Thus the local product structure is restricted to the sets U_α and U_β separately.

If M is a C^p manifold, we see at once that $T(M)$ is a C^{p-1} manifold, for the transformation in equation (1) uses up one derivative.

For use later we now define the projection map $\pi: T(M) \to M$.

<u>Definition</u>. The projection map $\pi: T(M) \to M$ is defined by

$$\pi(x^i, \lambda^i) = x^i$$

6.6 Vector Fields. We can now define a vector field on M in several ways.

Definition 1. If M is a C^p manifold, a C^{p-1} vector field on M is a C^{p-1} map $V: M \to T(M)$ such that $\pi \circ V: M \to M$ is the identity.

In local coordinates V is expressed by

$$V: x^i \to [x^i, \lambda^i(x)]$$

where $\lambda^i(x)$ are C^{p-1} functions. For consistency in the overlap of local coordinates the two maps

$$x^i \to [x^i, \lambda^i(x)] \quad \text{and} \quad \bar{x}^i = [\bar{x}^i, \lambda^i(\bar{x})]$$

must be such that

(1) x and $\bar{x}$ represent the same point P

(2) $\bar{\lambda}^i [\bar{x}(x)] = \lambda^i(x) \dfrac{\partial \bar{x}^j}{\partial x^i}$

Definition 2. If M is a C^∞ or a C^ω manifold, a C^p vector field X is a C^p map of the C^∞(or C^ω) functions on M into the C^∞(or C^ω) functions on M such that:

(1) $X\,[a(P)f + b(P)g]_P = a(P)(Xf)_P + b(P)(Xg)$ for all $P \varepsilon M$.

That is: X is linear over the C^∞(or C^ω) functions $a(P)$ and $b(P)$.

(2) $X\,(fg)_P = f(P)(Xg)_P + g(P)(Xf)_P$ for all $P \varepsilon M$.

There is an obvious generalization of this definition to the $C^p(1 \leq p < \infty)$ case following the lines of our definition of $X_P f$ in Section 6.4.

6.7 Lie Bracket. In the above terminology we can now give a smooth definition of the Lie Bracket (Sec. 3.11) $[X_\alpha, X_\beta]$ in the C^∞ and $C^{(\omega)}$ cases.

Definition. Let X_α and X_β be two C^∞(or $C^{(\omega)}$) vector fields on M. Then $[X_\alpha, X_\beta]f$ is a mapping of the C^∞(or $C^{(\omega)}$) functions on M into the C^∞(or $C^{(\omega)}$) functions on M defined by:

$$[X_\alpha, X_\beta]f = X_\alpha(X_\beta f) - X_\beta(X_\alpha f)$$

It is an exercise to show that according to this definition $[X_\alpha, X_\beta]$ is a vector field on M.

6.8 Cotangent Vectors on M. If we begin with the vector space of tangent vectors T_PM, we can form the space of dual vectors, $(T_PM)^*$. We shall call these cotangent vectors. We shall write the canonical basis for $(T_PM)^*$ as $(dx^1,\dots,dx^n)$. Thus

$$dx^i\left(\frac{\partial}{\partial x^i}\right)_P = \delta^i_j$$

There is a cotangent vector, df, associated with each C^∞ (or $C^{(\omega)}$) function f defined near P on M. It is defined by the equality;

$$df(X_P) = X_P(f)$$

where X_P is a tangent vector.

Theorem 5. In terms of the canonical basis of cotangent vectors

$$df = \left(\frac{\partial f}{\partial x^i}\right)_P dx^i$$

Proof. By definition

$$df\left(\frac{\partial}{\partial x^i}\right) = \left(\frac{\partial}{\partial x^i}\right)_P f = \left(\frac{\partial f}{\partial x^i}\right)_P$$

Also

$$\left[\left(\frac{\partial f}{\partial x^j}\right)_P dx^j\right]\left[\frac{\partial}{\partial x^i}\right] = \left(\frac{\partial f}{\partial x^j}\right)_P \delta^j_i = \left(\frac{\partial f}{\partial x^i}\right)_P$$

Since the two cotangent vectors df and $\left(\frac{\partial f}{\partial x^i}\right)_P dx^i$ have the same values on the set of basis vectors of $T_P(M)$, they are equal.

Remark. The expression df is often called the differential of f, but it is not the same as the df defined in Section 1.5. Here df is a cotangent vector; in 1.5 df is a linear transformation. A close relationship between the two definitions exists, however, because of Theorem 5. A better name for the df of the present section is the exterior derivative of f.

In terms of the canonical basis we shall write an arbitrary cotangent vector in the notation

$$\omega = \omega_i dx^i$$

where ω_i are the components of ω.

As in the case of tangent vectors, we can define the bundle of cotangent vectors $T(M)^*$. Again this is a local product space and is a C^{p-1} manifold if M is a C^p manifold. Let a coordinate system of $U_\alpha \times R^n$ be $[x^i, \omega_i(x)]$ where $\omega_i(x)$ are components of a cotangent vector at x. In overlapping coordinate systems we have

$$\omega_i(x) = \bar{\omega}_j[\bar{x}(x)] \frac{\partial \bar{x}_j}{\partial x^i}$$

in accordance with Theorem 3 of Chapter 5.

Thus we can define a field of cotangent vectors as a C^{p-1} map: $M \to T(M)^*$ as before. These have a special name:

<u>Definition</u>. A differential 1-form is a field of cotangent vectors. The notation is $\omega = \omega_i(x)dx^i$.

Because of its definition we can evaluate a 1-form on a vector field. Thus

$$[\omega(X)]_P = \omega_P(X_P) \quad \text{at each point} \quad P.$$

Thus $\omega(X)$ is a function, which is C^∞ (or C^ω) if ω and X are C^∞ (or C^ω). It is easy to show that ω is "linear over the functions"; i.e.

$$\omega(fX + gY) = f\omega(X) + g\omega(Y)$$

where f and g are functions.

If $X = \lambda^i(x) \frac{\partial}{\partial x^i}$ and $\omega = \omega_j(x)dx^j$ in local coordinates,

$$\begin{aligned} \omega(X) &= \left[\omega_j(x)dx^j\right]\left[\lambda^i(x) \frac{\partial}{\partial x^i}\right] \\ &= \omega_j(x)\lambda^i(x)\delta^j_i \\ &= \omega_i(x)\lambda^i(x) \end{aligned}$$

One-forms occur traditionally in the theory of <u>total differential</u> equations in R^3. In this theory we are asked to solve the total differential equation:

$$Pdx + Qdy + Rdz = 0.$$

A solution is a two-dimensional submanifold to which the vector (P,Q,R) is orthogonal. See section 3.12. In the present terminology $\omega = Pdx + Qdy + Rdz$ is a 1-form and we are seeking a two-dimensional submanifold on which it is zero. This means that $\omega(X) = 0$ for all vectors X tangent to the submanifold. If such a vector X has components (X,Y,Z) then

$$\omega(X) = (P,Q,R) \cdot (X,Y,Z).$$

If this is to be zero for all tangent vectors, it follows that (P,Q,R) is orthogonal to the submanifold. So from 3.12 there exists a two-dimensional submanifold on which $Pdx + Qdy + Rdz = 0$ iff

$$(P,Q,R) \cdot \text{curl } (P,Q,R) = 0$$

6.9 <u>Tensor Fields and p-forms on M</u>. The definitions of these objects are completely analogous to those of vector fields and 1-forms.

To define a tensor field of type (r,s) we form the tensor bundle $T_{r,s}(M)$. This is again a manifold whose open sets are $U_\alpha \times R^{n^{r+s}}$. The local coordinates of a point in this manifold are $\left[x^i,\ T^{q_1\cdots q_r}_{p_1\cdots p_s}(x)\right]$ so that a point in the manifold is a pair consisting of a point P of M and the components of a tensor of type (r,s) at P. A tensor field is then a map: $M \to T_{r,s}(M)$ having the required degree of differentiability.

Similarly a p-form is a field of p-covectors defined in the same way. In local coordinates we shall write p-forms in the notation:

$$\omega^p = \frac{1}{p!}\,\omega_{i_1\cdots i_p}(x)\;dx^{i_1}\wedge\cdots\wedge dx^{i_p}$$

where $\omega_{i_1\cdots i_p}$ is alternating. As in Section 5.8, we can then define the exterior product $\omega^p \wedge \omega^q$ to be a $(p + q)$-form.

Remark. You will observe that p-forms are already familiar as integrands in Advanced Calculus. For example:

$Pdx + Qdy + Rdz$ is a 1-form which is the integrand in a line integral.

$Pdy \wedge dz + Qdz \wedge dx + Rdx \wedge dy$ is a 2-form which is the integrand in a surface integral.

It will be convenient, hereafter, to consider functions on M to be 0-forms.

6.10 Exterior Derivatives of p-forms on M. First we recall the definition of the exterior derivative of a 0-form (a function) in Section 6.8. In local coordinates

$$df = \frac{\partial f}{\partial x^i}\,dx^i$$

We generalize this to general p-forms as follows:

Definition (local). The exterior derivative of a p-form

$$\omega = \frac{1}{p!}\omega_{i_1\cdots i_p}(x)\;dx^{i_1}\wedge\cdots\wedge dx^{i_p}$$

is

$$d\omega = \frac{1}{p!}\,d\omega_{i_1\cdots i_p}\wedge dx^{i_1}\wedge\cdots\wedge dx^{i_p}$$

where, of course,

$$d\omega_{i_1\ldots i_p} = \frac{\left(\partial\, \omega_{i_1\ldots i_p}\right)}{\partial x^{i_{p+1}}}\, dx^{i_{p+1}}$$

Examples.

1. If $\omega = Pdx + Qdy$ in R^2

$$d\omega = \left(\frac{\partial P}{\partial x} dx + \frac{\partial P}{\partial y} dy\right) \wedge dx + \left(\frac{\partial Q}{\partial x} dx + \frac{\partial Q}{\partial y} dy\right) \wedge dy$$

$$= \left(\frac{\partial Q}{\partial x} - \frac{\partial P}{\partial y}\right) dx \wedge dy$$

2. If $\omega = P\, dy \wedge dz + Q\, dz \wedge dx + R\, dx \wedge dy$ in R^3

$$d\omega = \left(\frac{\partial P}{\partial x} + \frac{\partial Q}{\partial y} + \frac{\partial R}{\partial z}\right) dx \wedge dy \wedge dz$$

3. If ω is an n-form in R^n, $d\omega = 0$. We leave as exercises the important results:

Theorem 6.

(a) $d\, d\omega = 0$

(b) $d(\omega^p \wedge \pi^q) = d\omega^p \wedge \pi^q + (-1)^p \omega^p \wedge d\pi^q$

Definition (Abstract). The exterior derivative d is a map from the p-forms to the (p+1)-forms such that:

(1) df is an defined in Section 6.8.

(2) $d(\omega + \pi) = d\omega + d\pi$

(3) $d(\omega^p \wedge \pi^q) = d\omega^p \wedge \pi^q + (-1)^p \omega^p \wedge d\pi^q$

(4) $d\, d\, f = 0$ for functions f.

We leave it to you to prove that the abstract definition implies the local definition.

A third "invariant" definition of $d\omega$ is the following. We give it only for a 1-form.

<u>Definition (Invariant)</u>. If ω is a one-form

$$d\omega(X,Y) = X[\omega(Y)] - Y[\omega(X)] - \omega[X,Y].$$

To prove that this agrees with the local definition, use components.

Let $$\omega = a_i dx^i, \quad X = X_1^i \frac{\partial}{\partial x^i}, \quad Y = X_2^i \frac{\partial}{\partial x^i}$$

Then $$\omega(Y) = (a_i dx^i)\left(X_2^j \frac{\partial}{\partial x^j}\right) = a_i X_2^i$$

and $$X\omega(Y) = X_1^j \frac{\partial}{\partial x^j}\left(a_i X_2^i\right)$$

$$= X_1^j X_2^i \frac{\partial a_i}{\partial x^j} + a_i X_1^j \frac{\partial}{\partial x^j}\left(X_2^i\right)$$

Similarly

$$Y\omega(X) = X_1^j X_2^i \frac{\partial a_j}{\partial x^i} + a_i X_2^j \frac{\partial}{\partial x^j}\left(X_1^i\right)$$

So

$$X\omega(Y) - Y\omega(X) = X_1^j X_2^i \left(\frac{\partial a_i}{\partial x^j} - \frac{\partial a_j}{\partial x^i}\right) + a_i [X,Y]^i$$

and $$X\omega(Y) - Y\omega(X) - \omega[X,Y]$$

$$(1) \qquad = X_1^j X_2^i \left(\frac{\partial a_i}{\partial x^j} - \frac{\partial a_j}{\partial x^i}\right)$$

But in the local form:

$$d\omega = \frac{\partial a_i}{\partial x^j} dx^j \wedge dx^i$$

$$d\omega[X,Y] = \frac{\partial a_i}{\partial x^j} [dx^j(X)dx^i(Y) - dx^j(Y)dx^i(X)]$$

$$= \frac{\partial a_i}{\partial x^j} \{X_1^j X_2^i - X_2^j X_1^i\}$$

$$(2) \qquad = \left[\frac{\partial a_i}{\partial x^j} - \frac{\partial a_j}{\partial x^i}\right] X_1^j X_2^i$$

Since (1) and (2) are equal, the result is proved.

6.11 <u>Frobenius Theorem (Third Form)</u>. Let ω^a $(a = r+1 \ldots n)$ be $n-r$ linearly independent 1-forms in $U \subset R^n$ where $r \geq 2$. The system $\omega^a = 0$ is called a system of Total Differential Equations. A solution of this system is an r-dimensional submanifold, M^r, on which $\omega^a = 0$.

To describe another interpretation, let

$$\omega^a = P_i^a \, dx^i$$

Then $\omega^a = 0$ on M^r means that $\omega^a\left(\frac{\partial X}{\partial u^\alpha}\right) = 0$ for all $\frac{\partial X}{\partial u^\alpha}$ tangent to M^r. This is equivalent to $P_i^a \frac{\partial X^i}{\partial u^\alpha} = 0$, and says that P_i^a are normal to M^r.

<u>Theorem 7</u>. The system $\omega^a = 0$ has a solution M^r through each

point of U iff there exists a matrix of 1-forms λ_b^a such that in U,

$$d\omega^a = \lambda_b^a \wedge \omega^b$$

<u>Proof</u>. Complete the given system to a basis for 1-forms in U: $\{\omega^\alpha, \omega^a\}$ $\alpha = 1\ldots r$, $a = r+1\ldots n$. Then there are vector fields X_α and X_a such that $\omega^\alpha(X_\beta) = \delta_\beta^\alpha, \omega^a(X_\alpha) = 0$, $\omega^a(X_b) = \delta_b^a$ and $\omega^\alpha(X_b) = 0$.

In 6.10 it was shown that

$$d\omega^a(X_\alpha, X_\beta) = X_\alpha[\omega^a(X_\beta)] - X_\beta[\omega^a(X_\alpha)] - \omega^a[X_\alpha, X_\beta]$$

$$= -\omega^a[X_\alpha, X_\beta]$$

If $d\omega^a = \lambda_b^a \wedge \omega^b$, then $d\omega^a(X_\alpha, X_\beta) = 0$.

Hence $\omega^a[X_\alpha, X_\beta] = 0$. Since $[X_\alpha, X_\beta]$ is a vector field, we can write it in the form:

$$[X_\alpha, X_\beta] = C_{\alpha\beta}^a X_a + C_{\alpha\beta}^\gamma X_\gamma$$

Hence $C_{\alpha\beta}^a = \omega^a[X_\alpha, X_\beta] = 0$ and $[X_\alpha, X_\beta] = C_{\alpha\beta}^\gamma X_\gamma$.

By the second form of the Frobenius Theorem, there is then a manifold M^r to which X_α are tangent, and hence on which $\omega^a = 0$.

Conversely, suppose that M^r is a solution of $\omega^a = 0$. Let Y_α be a basis for the tangent vectors to M^r. Then $\omega^a(Y_\beta) = 0$ and so

$$d\omega^a(Y_\alpha, Y_\beta) = -\omega^a[Y_\alpha, Y_\beta] = -\omega^a C^\gamma_{\alpha\beta} Y_\gamma = 0$$

Now in general

$$d\omega^a = A^a_{bc}\omega^c \wedge \omega^b + B^a_{\alpha b}\omega^\alpha \wedge \omega^b + C^a_{\alpha\beta}\omega^\alpha \wedge \omega^\beta$$

Since $d\omega^a(Y_\alpha, Y_\beta) = 0$ and $\omega^a(Y_\beta) = 0$, it follows that $C^a_{\alpha\beta} = 0$.

Hence $d\omega^a = (A^a_{bc}\omega^c + B^a_{\alpha b}\omega^\alpha) \wedge \omega^b = \lambda^a_b \wedge \omega^b$ where λ^a_b is a matrix of one-forms.

The requirement that $d\omega^a = \lambda^a_b \wedge \omega^b$ can be stated in several equivalent forms. Let $\Omega = \omega^{r+1} \wedge \ldots \wedge \omega^n$. Then we have the theorem:

<u>Theorem 8</u>. The following statements are equivalent:

(1) $d\omega^a = \lambda^a_b \wedge \omega^b$

(2) $d\Omega = \lambda \wedge \Omega$ where λ is a one-form.

(3) $d\omega^a \wedge \Omega = 0$

<u>Proof</u>. (1) $\rightarrow$ (2)

$$\begin{aligned} d\Omega = \; & d\omega^{r+1} \wedge \omega^{r+2} \wedge \ldots \wedge \omega^n \\ & - d\omega^{r+2} \wedge \omega^{r+1} \wedge \omega^{r+3} \wedge \ldots \wedge \omega^n \\ & + \text{etc.} \\ = \; & \lambda^{r+1}_{r+1} \wedge \omega^{r+1} \wedge \ldots \wedge \omega^n \\ & - \lambda^{r+2}_{r+2} \wedge \omega^{r+2} \wedge \omega^{r+1} \wedge \omega^{r+3} \wedge \ldots \wedge \omega^n \\ & + \text{etc.} \end{aligned}$$

$$= \lambda^{r+1}_{r+1} \wedge \Omega + \lambda^{r+2}_{r+2} \wedge \Omega + \ldots + \lambda^{n}_{n} \wedge \Omega$$

$$= \lambda \wedge \Omega \quad \text{where} \quad \lambda = \lambda^{r+1}_{r+1} + \ldots + \lambda^{n}_{n} \quad \text{is a one-form.}$$

(2) → (3)

Since $\Omega \wedge \omega^a = 0$

$$0 = d(\Omega \wedge \omega^a) = d\Omega \wedge \omega^a \pm \Omega \wedge d\omega^a$$

But $d\Omega \wedge \omega^a = \lambda \wedge \Omega \wedge \omega^a = 0$

Hence $d\omega^a \wedge \Omega = 0$

(3) → (1)

$$d\omega^a = A^a_{bc}\omega^c \wedge \omega^b + B^a_{\alpha b}\omega^\alpha \wedge \omega^b + C^a_{\alpha\beta}\omega^\alpha \wedge \omega^\beta$$

$$0 = d\omega^a \wedge \Omega = 0 + 0 + C^a_{\alpha\beta}\omega^\alpha \wedge \omega^\beta \wedge \Omega$$

Hence $C^a_{\alpha\beta} = 0$

And $d\omega^a = (A^a_{bc}\omega^c + B^a_{\alpha b}\omega^\alpha) \wedge \omega^b = \lambda^a_b \wedge \omega^b$

Example. Suppose we have a single form ω. Then the equation $\omega = 0$ has a solution iff $d\omega \wedge \omega = 0$ from (3).

If $n = 3$, $r = 2$, let $\omega = P dx + Q dy + R dz$

Then

$$d\omega \wedge \omega = \left[P\left(\frac{\partial R}{\partial y} - \frac{\partial Q}{\partial z}\right) + Q\left(\frac{\partial P}{\partial y} - \frac{\partial R}{\partial x}\right) + R\left(\frac{\partial Q}{\partial x} - \frac{\partial P}{\partial y}\right)\right] dx \wedge dy \wedge dz$$

Thus the condition for a solution of

$$P dx + Q dy + R dz = 0$$

is that $\underset{\sim}{A} \cdot \text{curl}\ \underset{\sim}{A} = 0$ where $\underset{\sim}{A} = (P, Q, R)$.

6.12 Mappings. In this section we shall examine how functions, vectors, and forms are transformed when we map one manifold into (or onto) another. Let M and N be differentiable manifolds and

$$\phi: M \to N$$

a C^∞ map. That is, the coordinate functions from a local coordinate system in M to a local coordinate system in N are C^∞.

(1) Let f be a C^∞ function: $N \to R$. Then $\phi^* f$ is a function $M \to R$ defined by

$$(\phi^* f)(P) = f \circ \phi(P) \quad \text{where } P \in M$$

(2) Let X be a vector at P on M. Then $\phi_* X$ is a vector at $\phi(P)$ on N defined by:

$$[\phi_* X)f]_{\phi(P)} = X(\phi^* f)_P$$

where f is a function: $N \to R$.

This mapping is meaningful for vector fields only if ϕ is one-to-one, but is not well-defined otherwise.

ϕ_* thus maps $T_P(M)$ into $T_{\phi(P)}N$. If x^i are local coordinates on M at P and y^α are local coordinates on N at $\phi(P)$, then ϕ is defined locally by $y^\alpha = \phi^\alpha(x)$, where $\phi^\alpha(x)$ are C^∞ functions. If $\lambda^i(x)$ are components of X_P and $\xi^\alpha(y)$ are components of $(\phi_* X)_{\phi(P)}$ then

$$\lambda^i(x) \frac{\partial y^\alpha}{\partial x^i} = \xi^\alpha[y(x)]$$

Hence if $\left(\frac{\partial y^\alpha}{\partial x^i}\right)$ is a square, non-singular matrix, ϕ_* is one-to-one. And if rank $\left(\frac{\partial y^\alpha}{\partial x^i}\right) = \dim N$, ϕ_* is onto.

The map $\phi_* T_P(M) \to T_{\phi(P)}N$ is often written $d\phi$ and is called the <u>differential</u> of ϕ. This agrees with the terminology of Section 1.5. In this sense $d\phi$ is a linear approximation to ϕ near P.

(3) Let ω be a p-covector at $\phi(P)$ on N. Then $\phi^*\omega$ is a p-covector at P on M defined by

$$\phi^*\omega(X_1,\ldots,X_p)_P = \omega(\phi_* X_1,\ldots,\phi_* X_p)_{\phi(P)}$$

This applies to p-forms as well for any ϕ.

<u>Theorem 9</u>. $d(\phi^* f) = \phi^*(df)$

<u>Proof</u>. (1) $d(\phi^* f)(X) = X(\phi^* f)$ by definition.

(2) $(\phi^* df)X = df(\phi_* X) = (\phi_* X)f = X(\phi^* f)$

<u>Lemma</u>. $\phi^*(\omega \wedge \pi) = (\phi^*\omega) \wedge (\phi^*\pi)$

Proof left to an exercise.

<u>Theorem 10</u>. $d(\phi^*\omega) = \phi^* d\omega$ for all ω.

<u>Proof</u>. Let $\omega = g(y)\, dy^1 \wedge \ldots \wedge dy^p$ where y^i are local coordinates on N. Then

(1) $d\omega = dg \wedge dy^1 \wedge \ldots \wedge dy^p$

$$\phi^* d\omega = \phi^* dg \wedge \phi^* dy^1 \wedge \ldots \wedge \phi^* dy^p$$

$$= d(\phi^* g) \wedge d(\phi^* y^1) \wedge \ldots \wedge d(\phi^* y^p)$$

$$(2)\quad \emptyset^*\omega = (\emptyset^*g)\cdot \emptyset^*dy^1\wedge\ldots\wedge\emptyset^*dy^p$$

$$= (\emptyset^*g)\cdot d(\emptyset^*y^1)\wedge\ldots\wedge d(\emptyset^*y^p)$$

$$d(\emptyset^*\omega) = d(\emptyset^*g)\wedge d(\emptyset^*y^1)\wedge\ldots\wedge d(\emptyset^*y^p)$$

6.13 Sard's Theorem. This theorem is one of great utility in the discussion of maps: $\emptyset: M \to N$. Since it holds for C^1 maps, we assume in this section that M and N are C^1 manifolds.

Theorem 11 (Sard). Let M and N be C^1 differentiable manifolds of the same dimensions and let $\emptyset: M \to N$ be a C^1 map. Let $A \subset M$ be the set of critical points of $\emptyset$ on M. Then the set of critical values, $\emptyset(A) \subset N$ has measure zero.

To understand this theorem, you must apply two definitions:

Definitions. A critical point $P \in M$ of the map $\emptyset$ is a point at which $d\emptyset$ is singular; i.e, the Jacobian $|\partial y^\alpha/\partial x^i|$ of Section 6.12 is equal to zero. If P is a critical point of $\emptyset$, then $\emptyset(P) \in N$ is a critical value. A point of M which is not critical is called regular.

Definition. A subset S of N has measure zero iff:

(1) $S = \cup S_\alpha$ where each S_α lies in a coordinate neighborhood $U_\alpha \subset N$. We assume one-to-one maps $f_\alpha: U_\alpha \to I^n$ where I^n is the interior of the unit cube in R^n.

(2) $f_\alpha(S_\alpha)$ has measure zero in R^n

Sard's theorem then follows from the following result:

Lemma. Let $g: I^n \to I^n$ be a C^1 map. Then the set of critical values of g has measure zero.

Proof. 1. We may divide I^n into L^n cubes $C(L)$ each of side $1/L$.

2. Since g is differentiable, it follows that for given $\varepsilon > 0$, we can find an L such that for all x,y in each $C(L)$:

$$|g(y) - g(x) - (dg)_x(y - x)| < \varepsilon\, |y - x| \leq \frac{\varepsilon\sqrt{n}}{L}$$

3. Let A be the set of critical points of g, and suppose that $x \in C \cap A$. Then $\det\,(dg)_x = 0$.

Hence

$\{g(x) + (dg)_x\,(y - x)\,|\,y \in C,\ x \text{ fixed}\}$ lies in an $n-1$ dimensional hyperplane V of I^n passing through $g(x)$. Therefore, from (2), $g(y)$ lies within $\frac{\varepsilon\sqrt{n}}{L}$ of V.

4. Also there is a Lipschitz constant K such that

$|g(y) - g(x)| < K\,|y - x| \leq \frac{K\sqrt{n}}{L}$ for x, y in any cube $C(L)$.

5. From (3) and (4), if C intersects A, the set $\{g(y)\,|\,y \in C\}$ is contained in a cylinder whose height is less than $\frac{2\,\varepsilon\,\sqrt{n}}{L}$ and whose base is a $(n-1)$ sphere centered at $g(x)$ whose radius is less than $\frac{K\,\sqrt{n}}{L}$.

6. The volume of this cylinder is less than $\frac{b\varepsilon}{L^n}$ where b is a constant depending only on K and n.

7. There are at most L^n cubes C that intersect A, so A lies in a set whose volume is less than $b\,\varepsilon$. Since ε is arbitrary, A has measure zero.

Remark. This is the so-called "easy case" of Sard's Theorem. In the general case, M and N may have different dimensions, and the theorem is more complicated.

Sard's Theorem has the following consequences.

Definition. A C^1 map $\emptyset: M \rightarrow N$ where M and N have the same dimensions is called nondegenerate iff M has at least one regular point of $\emptyset$.

Theorem 12. If $\emptyset: M \rightarrow N$ is a nondegenerate map, then there is at least one point of N which is not a critical value of $\emptyset$.

Proof. By definition there is at least one regular point, r, of M. Then there is a neighborhood U(r) within which $\emptyset$ is one-to-one so that $\emptyset[U(r)]$ does not have measure zero. Hence from Sard's Theorem $\emptyset[U(r)]$ must contain points which are not critical values.

Theorem 13. If $\emptyset: M \rightarrow N$ is a nondegenerate map, and if M is compact, there is a point p of N whose set of preimages $\emptyset^{-1}(p)$ consists of at most a finite number of points.

Proof. 1. Choose $p \in N$ such that p is not a critical value of $\emptyset$. [Possible by Theorem 12].

2. For each regular point $r \in M$, choose a neighborhood U(r) in which $\emptyset$ is one-to-one.

3. If c is a critical point of $\emptyset$, it must be true that $\emptyset(c) \neq p$, for p is not a critical value. Hence for each c we may choose a neighborhood V(c) such that $\emptyset[V(c)]$ does not contain p.

4. The entire set of U(r), V(c) cover M. Since M is compact, a finite subset of these, say $U(r_i)$, $V(c_j)$ covers M.

5. Since p can only be contained in the images of $U(r_i)$ and since in each of these $\emptyset$ is one-to-one, there are only a finite number of points, x, in M for which $\emptyset(x) = p$.

6.14 Connections and Covariant Derivatives. In this section, we give an abstract definition of a connection and a covariant derivative which generalizes the special definition given in Chapter 4.

Definition. Let f be a C^∞ function and X,Y,Z be C^∞ vector fields on M. The directional covariant derivative is an operator ∇_X such that:

(1) $\nabla_X f = X(f)$

(2) $\nabla_X Y = Z$

(3) $\nabla_X(Y + Z) = \nabla_X Y + \nabla_X Z$

(4) $\nabla_{X+Y}(Z) = \nabla_X Z + \nabla_Y Z$

[(3) and (4) state that $\nabla_X Y$ is bilinear in X and Y]

(5) $\nabla_{fX}(Y) = f\nabla_X Y$

(6) $\nabla_X(fY) = (\nabla_X f)Y + f\nabla_X Y$

If in particular $X = e_i$, $Y = e_j$ where e_i and e_j are local basis elements of $T(M)$, we define the kth component Z^k of Z to be:

$$Z^k = [\nabla_{e_i}(e_j)]^k = \Gamma^k_{ij}$$

where Γ^k_{ij} are n^3 C^∞ functions. If $X = X^i e_i$ and $Y = Y^i e_i$, it follows from properties (1) to (6) that

$$(7) \qquad (\nabla_X Y)^k = Y^k_{,i} X^i \quad \text{where} \quad Y^k_{,i} = \frac{\partial Y^i}{\partial x^i} + \Gamma^k_{ij} Y^j$$

Conversely, if local functions Γ^k_{ij} are given, formula (7) defines a $\nabla_X Y$ which has properties (2) to (6).

<u>Remarks</u>. (1) The Γ^k_{ij} are not necessarily symmetric in i and j as was true in Chapter 4. Nor are they required to be Christoffel symbols defined in terms of a metric tensor g_{ij}.

(2) The terms connection, covariant derivative, and directional covariant derivative are unfortunately used somewhat interchangeably in the literature. The following is the standard usage:

∇_X is the <u>directional covariant</u> derivative;

$Y^k_{,i}$ are the components of the covariant derivative of Y, in local coordinates;

$Y^k_{,i}$ are the components of a tensor of type (1,1);

Γ^k_{ij} are the components of the <u>connection</u> in local coordinates.

The operator ∇_X can be extended to 1-forms as follows:

<u>Definition</u>. If ω is a 1-form, then $\nabla_X \omega$ is a 1-form whose value on an arbitrary vector field Y is defined to be:

$$(\nabla_X \omega)Y = \nabla_X[\omega(Y)] - \omega(\nabla_X Y)$$

If in local coordinates $X = X^i \frac{\partial}{\partial x^i}$, $Y = Y^i \frac{\partial}{\partial x^i}$, and $\omega = \omega_i dx^i$, it is an easy calculation to show that

$$(8) \qquad (\nabla_X \omega)Y = \omega_{i,k} Y^i X^k \quad \text{where} \quad \omega_{i,k} = \frac{\partial \omega_i}{\partial x^k} - \Gamma^j_{ki} \omega_j$$

Then $\omega_{i,k}$ are the local components of the covariant derivative of ω, and are the components of a tensor of type (0,2).

Since any tensor field is a linear combination (over the functions) of the tensor products of vector fields and 1-forms, the operator ∇_X can be extended to map a tensor field of type (r,s) into another field of the same type. We illustrate for a field of type (1,1).

Let $T(1,1) = f^{\alpha\beta} Y_\alpha \otimes \omega_\beta$, then by definition

$$\nabla_X T = (\nabla_X f^{\alpha\beta})\, Y_\alpha \otimes \omega_\beta + f^{\alpha\beta}(\nabla_X Y_\alpha) \otimes \omega_\beta + f^{\alpha\beta} Y_\alpha \otimes \nabla_X \omega_\beta$$

In terms of covariant derivatives, this is equivalent to:

$$(9) \qquad (T^i_j)_{,k} = \frac{\partial f^{\alpha\beta}}{\partial x^k} Y^i_\alpha \, \omega_{\beta j} + f^{\alpha\beta} Y^i_{\alpha,k} \, \omega_{\beta j} + f^{\alpha\beta} Y^i_\alpha \, \omega_{\beta j,k}$$

If we use (7) and (8) this reduces to:

$$(10) \qquad T^i_{j,k} = \frac{\partial T^i_j}{\partial x^k} + \Gamma^i_{kr} T^r_j - \Gamma^r_{kj} T^i_r$$

which are the components of a tensor of type (1,2).

6.15 The Covariant Exterior Derivative. This is an operator on tensor-valued p-forms which are defined as follows.

Definition. A tensor-valued p-form is a tensor of type (r,p) such that:

$$T(r,p)(f^{i_1},\ldots,f^{i_r}, e_{j_1},\ldots,e_{j_u},\ldots,e_{j_v},\ldots,e_{j_s}) =$$

$$- T(r,p)(f^{i_1},\ldots,f^{i_r}, e_{j_1},\ldots,e_{j_v},\ldots,e_{j_u},\ldots,e_{j_s})$$

for each pair of basis elements e_{j_u} and e_{j_v}.

The exterior covariant derivative is an operator D such that D T(r,p) is a tensor-valued (p+1)-form T(r,p+1). By definition:

(1) $Df = df$ for functions f

(2) $D\omega = d\omega$ for any p-form

(3) If X and Y are vector fields, DY is the vector valued 1-form whose value on X is $(DY)X = \nabla_X Y$.

Then D can be extended to arbitrary tensor-valued p-forms.

For example if X is a vector field and ω is a 1-form, $X \otimes \omega$ is a vector-valued 1-form. Then

$$D(X \otimes \omega)(Y,Z) = (\nabla_Y X) \otimes \omega(Z) - (\nabla_Z X) \otimes \omega(Y) + X \otimes d\omega(Y,Z).$$

6.16 Parallel Displacement. We can also consider covariant differentiation along a curve. Let C be a differentiable curve on M whose parametric equations in local coordinates are $x^i = f^i(t)$. Then dx^i/dt is a tangent vector to the curve, which plays the role of the vector X in ∇_X. So from (7) we may consider

$$(\nabla_t Y)^k = Y^k_{,i} \frac{dx^i}{dt}$$

$$= \frac{dY^k}{dt} + \Gamma^k_{ij} \frac{dx^i}{dt} Y^j \tag{11}$$

This derivation assumes that Y is defined on M, but the formula (11) makes sense if Y is defined only on the curve C. So we have the definition:

Definition. If C is a curve on M with parameter t and Y is a vector field defined along C, then

$$(\nabla_t Y)^k = \frac{dY^k}{dt} + \Gamma^k_{ij} \frac{dx^i}{dt} Y^j \tag{11'}$$

This enables us to introduce the concept of a parallel field of vectors along a curve.

Definition. If C is a curve on M and Y is a vector field defined on C, then Y is parallel along C if $\nabla_t Y = 0$.

The equations $\nabla_t Y = 0$ are a system of ordinary differential equations which can be integrated if initial values of Y are given at $t = t_0$. So, if we have a curve C on M joining P and Q and a vector Y_P at P, the solution of $\nabla_t Y$ with initial values Y_P forms a parallel field on C. We say that Y_P has undergone a parallel displacement along C. In particular to each vector at P there is assigned a unique vector at Q relative to the curve C. Thus the process of parallel displacement along C provides a connection between T_PM and T_QM. This is the origin of the term connection in the context of this section.

Parallel displacement in general depends on the curve C, but it is independent of C if $Y^i_{,k} = 0$. These equations in general have solutions $Y^i(x)$ only if their integrability conditions are satisfied. These were computed in Exercise 7, Chapter 2. When solutions do exist, the space is said to be a space of distant parallelism, similar to Euclidean space.

Also when $Y^i = \frac{dx^i}{dt}$, we have

$$\left[\nabla_t\left(\frac{dx}{dt}\right)\right]^k = \frac{d^2x^k}{dt^2} + \Gamma^k_{ij}\frac{dx^i}{dt}\frac{dx^j}{dt} \tag{12}$$

When this is zero, we say that C is self-parallel or that C is a geodesic when t is arc-length s.

6.17 Orientation of M. Consider first a coordinate system (U_α, ϕ_α) on M. At each point $P \in U_\alpha$ we have the ordered basis for $T_P(M)$: $\left(\frac{\partial}{\partial x^1}, \ldots, \frac{\partial}{\partial x^n}\right)$. We choose this to be the "fundamental basis" of Section 5.9 and thus define an orientation on $T_P(M)$. If $(X_1(x), \ldots, X_n(x))$ is an ordered set of independent, continuous vector fields defined on U_α, their components $X^i_j(x)$ relative to these bases are continuous and so sign $\det(X^i_j)$ is constant on U_α. Thus, we have an orientation for the local tangent bundle $T_{U_\alpha}(M)$.

In two overlapping coordinate systems, U_α and U_β, the two bases $\left(\frac{\partial}{\partial x^1}, \ldots, \frac{\partial}{\partial x^n}\right)_{U_\alpha}$ and $\left(\frac{\partial}{\partial \bar{x}^1}, \ldots, \frac{\partial}{\partial \bar{x}^n}\right)_{U_\beta}$ are related in $U_\alpha \cap U_\beta$ by a linear transformation whose matrix is the

Jacobian, J, of $\psi_{\beta\alpha} = \phi_\beta \circ \phi_\alpha^{-1}$ (Section 6.1). If det J is positive, the orientations of $T_{U_\alpha}(M)$ and $T_{U_\beta}(M)$ agree.

If det J is positive for all pairs of overlapping coordinate systems on M, we can thus orient the whole of T(M). This orientation of T(M) is called an orientation of M itself. Hence we have the definitions:

Definitions. M is oriented if the Jacobian of each $\psi_{\beta\alpha}$ is positive. M is orientable if it can be covered with an oriented atlas.

Exercises

1. Prove that the set of tangent vectors X_p as defined in Section 6.3 forms a vector space.

2. Prove that the set of tangent vectors X_p as defined in Section 6.4 forms a vector space.

3. Prove that $[X_\alpha, X_\beta]$ as defined in Section 6.7 is a vector field.

4. Express $[X_\alpha, X_\beta]$ in terms of local coordinates in R^n and show that this agrees with our earlier definition in Section 3.11.

5. If on R^3, $\omega = Pdx + Qdy + Rdz$, find $d\omega$.

6. Prove $dd\omega = 0$ assuming the local definition for d.

7. Prove $d(\omega^p \wedge \pi^q) = d\omega^p \wedge \pi^q + (-1)^p \omega^p \wedge d\pi^q$ assuming the local definition of d.

8. Derive the local definition of $d\omega$ from the abstract definition assuming

$$\omega = \frac{1}{p!} \omega_{i_1 \ldots i_p}(x)\, dx^{i_1} \wedge \ldots \wedge dx^{i_p}$$

9. Prove $\phi^*(\omega \wedge \pi) = (\phi^*\omega) \wedge (\phi^*\pi)$

10. Prove that given Γ^k_{ij} and formula (7) of Section 6.14, $\nabla_X Y$ has properties (2) to (6).

11. Prove that $d\omega(X,Y) = (\nabla_Y \omega)X - (\nabla_X \omega)Y$

12. Prove that $[X_\alpha, X_\beta] = (\nabla_{X_\beta})X_\alpha - (\nabla_{X_\alpha})X_\beta$

13. Prove that $\omega_{i,k} = \frac{\partial \omega^i}{\partial X^k} - \Gamma^j_{ki}\,\omega_j$

14. Let $\omega = \omega_i dx^i$ be a 1-form on R^n, and let M be a submanifold of R^n with parametric equations $x^i = g^i(u^\alpha)$, which define a map $\emptyset : M \to R^n$. Show that

$$\emptyset^*\omega = \omega_i[x(u)]\,\frac{\partial x^i}{\partial u^\alpha}\,du^\alpha$$

15. As in exercise 14 let

$$\omega = \frac{1}{p!}\omega_{i_1 \dots i_p}\, dx^{i_1} \wedge \dots \wedge dx^{i_p}$$

Prove that

$$\emptyset^*\omega = \frac{1}{p!}\omega_{i_1 \dots i_p}[x(u)]\,\frac{\partial(x^{i_1} \dots x^{i_p})}{\partial(u^{\alpha_1} \dots u^{\alpha_p})}\,du^{\alpha_1} \wedge \dots \wedge du^{\alpha_p}$$

16. If now $M \subset R^n$ is a hypersurface with $\emptyset : M \to R^n$, and ω is an $(n-1)$ form on R^n,

$$\omega = (-1)^{i+1} P_i\, dx^1 \wedge \dots \wedge \widehat{dx^i} \wedge \dots \wedge dx^n$$

Then on M: $\emptyset^*\omega = P_i[x(u)]V^i = \langle P,N \rangle \sqrt{g}\, du^1 \wedge \dots \wedge du^{n-1}$

For notation see section 4.13.

17. Given a vector field X, the Lie Derivative, L_X, is a C^∞ operator such that $L_X T(r,s)$ is another tensor of the same type. It has an abstract definition from which the following formulas can be derived:

(1) $L_X f = Xf = (df)X$ for functions f.

(2) $L_X Y = [X,Y]$ for vector fields Y

(3) $(L_X \omega)Y = L_X[\omega(Y)] - \omega(L_X Y)$ for 1-forms ω

(4) $L_X(\omega_1 \wedge \omega_2) = (L_X\omega_1) \wedge \omega_2 + \omega_1 \wedge L_X(\omega_2)$ for any two forms.

Recall the definition of $i(X)$ in exercise 5 of Chapter 5. Then show that for 1-forms ω:

(*) $L_X\omega = i(X)\, d\omega + d[i(X)\omega]$

Hint: Evaluate both sides on a vector field Y and use the "invariant" definition of $d\omega$ in Section 6.10.

Then show that * is true if $\omega = \omega_1 \wedge \omega_2$ for 1-forms ω_1 and ω_2, and by induction that (*) holds for any p-form ω.

18. Prove that $H = L(X) - i(X)D$ is a linear transformation on $T_P(M)$.

19. Prove that $dL_X = L_X d$ and $DL_X = L_X D$

20. Prove that $DH = i(X)D^2$

21. Consider the spherical image of M^{n-1} in R^n described in Section 4.14. From the formula for $\Delta S = K_T \Delta V$ prove that

21. (Continued): ΔS is positive if the orientation of the image is the same as the orientation of S^{n-1} and negative otherwise.

22. Prove Theorem 7 (Third form of the theorem of Frobenius) directly from Theorem 8 of Chapter 1 (First form of the Frobenius theorem) using the following outline.

(1) Without loss of generality (prove this) we can choose

$$\omega^a = dx^a + f^a_\alpha(x^a, x^\alpha)dx^\alpha$$

(2) $$d\omega^a = \frac{\partial f^a_\alpha}{\partial x^b} dx^b \wedge dx^\alpha + \frac{\partial f^a_\alpha}{\partial x^\beta} dx^\beta \wedge dx^\alpha$$

(3) By hypothesis

$$d\omega^a = (\lambda^a_{bc}\, dx^c + \lambda^a_{b\beta}\, dx^\beta) \wedge (dx^b + f^b_\alpha\, dx^\alpha)$$

So $$\lambda^a_{bc} = 0; \qquad \frac{\partial f^a_\alpha}{\partial x^b} = -\lambda^a_{b\alpha}$$

$$\frac{\partial f^a_\alpha}{\partial x^\beta} - \frac{\partial f^a_\beta}{\partial x^\alpha} = -f^b_\alpha\, \lambda^a_{b\beta} + f^b_\beta\, \lambda^a_{b\alpha}$$

(4) Hence $$\frac{\partial f^a_\alpha}{\partial x^\beta} - \frac{\partial f^a_\beta}{\partial x^a} = f^b_\alpha \frac{\partial f^a_\beta}{\partial x^b} - f^b_\beta \frac{\partial f^a_\alpha}{\partial x^b}$$

(5) From Theorem 8, Chapter 1, there is a solution of

$$\frac{\partial x^a}{\partial x^\alpha} = f^a_\alpha$$

namely, $x^a + g^a(x^\alpha) = c^a$

which defines a submanifold M^r of R^n, such that

$$dx^a + f^a_\alpha\, dx^\alpha = 0$$

22. (Continued)

(6) Conversely if

$$\frac{\partial f^a_\alpha}{\partial x^\beta} - \frac{\partial f^a_\beta}{\partial\partial x^\alpha} = f^b_\alpha \frac{\partial f^a}{\partial x^b} - f^b_\beta \frac{\partial f^a_\alpha}{\partial x^b}$$

then

$$\lambda^a_{bc} = 0, \quad \lambda^a_{b\alpha} = -\frac{\partial f^a_\alpha}{\partial x^b}$$

are such that

$$d\omega^a = (\lambda^a_{bc}\, dx^c + \lambda^a_{b\beta}\, dx^\beta) \wedge \omega^b$$

$$= \lambda^a_b \wedge \omega^b$$

23. Given two manifolds M^n and N^p and a local basis for the one-forms on M: ω^i $(i = 1,\ldots,n)$. Also suppose that α^i $(i = 1,\ldots,n)$ are local one-forms on N. Find conditions sufficient for the existence of a local map $\phi: N \to M$ such that $\phi^*\omega^i = \alpha^i$. Is ϕ unique?

Chapter 7

Integration of Forms on Manifolds

7.1 Integration of Forms in R^p.

(1) Euclidean Cube. Let I^p be an open unit cube in R^p. One vertex is at the origin and the others are the unit points on the axes. Let I^p have the coordinate system $x^1,\ldots,x^p$ each having the range $0 < x^i < 1$ for $i = 1\ldots p$. The orientation of I^p is that given by the usual basis $e_1,\ldots,e_p$.

(2) Boundary of I^p. Let I_i^ε be the open $(p-1)$-dimensional face of I^p on which

$$x^i = \varepsilon = \begin{cases} 0 \\ 1 \end{cases}$$

The orientation of I_i^ε is given by the ordered set of vectors $(e_1,\ldots,\widehat{e_i},\ldots,e_p)$ where e_i is to be omitted.

Then the boundary ∂I^p is defined to be

$$\partial I^p = \sum_{i,\varepsilon} (-1)^{i+\varepsilon} I_i^\varepsilon$$

where $(-1)I_i^\varepsilon$ is I_i^ε with the orientation opposite to its given orientation. This is a formal sum which defines ∂I^p as a chain (see Section 7.3).

Observe that $N = (-1)^{\varepsilon+1}e_i$ is the outward normal to the face I_i^ε, and that the ordered set $(N,e_1,\ldots,\widehat{e_i},\ldots,e_p)$ has the orientation $(-1)^{i+\varepsilon}$ relative to $(e_1,\ldots,e_p)$. This motivates our choice of signs in the definition of ∂I^p.

(3) Integral of ω^p over I^p. Let $(dx^{i_1},\ldots,dx^{i_p})$ be an

ordered basis of the one-forms on R^p with an orientation μ relative to $(e_1,\ldots,e_p)$. And let $\omega = f(x)dx^{i_1}\wedge\ldots\wedge dx^{i_p}$ be a p-form defined on an open set of R^p containing I^p. Then by definition

$$\int_{I^p} \omega^p = \mu\int_{I^p} f(x)dx^1\ldots dx^p$$

where the right side is interpreted as a Riemann integral. Thus, we have defined an <u>oriented</u> integral whose sign depends on the order in which the differentials appear.

(4) <u>Integral of ω^{p-1} over ∂I^p</u>. Let

$$\omega^{p-1} = \frac{1}{(p-1)!}\,\omega_{i_1\ldots i_{p-1}}\,dx^{i_1}\wedge\ldots\wedge dx^{i_{p-1}}$$

(where $i_1,\ldots,i_p$ range over $1\ldots p$) be defined on an open set of R^p containing I^p.

Let $i\colon \partial I_i^\varepsilon \to R^p$ be the inclusion map. Then

$$i^*\omega_{p-1}(I_i^\varepsilon) = \omega_{1\ldots\hat{i}\ldots p}(x^1,\ldots,(x^i = \varepsilon),\ldots x^p)dx^1\wedge\ldots\wedge\widehat{dx^i}\wedge\ldots\wedge dx^p$$

So as in (3) $\int_{I_i^\varepsilon} i^*\omega_{p-1}$ is defined. Finally we define

$$\int_{\partial I^p} i^*\omega^{p-1} = \sum_{i,\varepsilon} (-1)^{i+\varepsilon}\int_{I_i^\varepsilon} i^*\omega_{p-1}$$

7.2 <u>Stokes' Theorem for I^p</u>.

<u>Theorem 1</u>. (Stokes) $\displaystyle\int_{\partial I^p} i^*\omega^{p-1} = \int_{I^p} d\omega^{p-1}$

<u>Proof</u>. If $\omega^{p-1} = \frac{1}{(p-1)!}\omega_{i_1\ldots i_{p-1}} dx^{i_1}\wedge\ldots\wedge dx^{i_{p-1}}$

then

$$d\omega^{p-1} = \left[\sum_i (-1)^{i-1} \frac{\partial\omega_{1\ldots\hat{i}\ldots p}}{\partial x^i}\right] dx^1\wedge\ldots\wedge dx^p$$

So $$\int_{I^p} d\omega^{p-1} = \int_{I^p}\left(\frac{\partial\omega_{2\ldots p}}{\partial x^1} - \frac{\partial\omega_{13\ldots p}}{\partial x^2} + \ldots\right) dx^1\ldots dx^p$$

where the right side is now a Riemann integral. We begin the evaluation of the integral of the first term in the sum by integrating with respect to x^1 from 0 to 1. The result is:

$$\omega_{2\ldots p}(1,x^2,\ldots,x^p) - \omega_{2\ldots p}(0,x^2,\ldots,x^p) =$$

$$\sum_\varepsilon (-1)^{1+\varepsilon}\omega_{2\ldots p}[(x^1=\varepsilon),x^2,\ldots,x^p]$$

Similarly we evaluate the second term by integrating with respect to x^2. The result is

$$-\sum_\varepsilon (-1)^{1+\varepsilon}\omega_{13\ldots p}[x^1,(x^2=\varepsilon),x^3,\ldots,x^p]$$

Hence $\int_{I^p} d\omega^{p-1} =$

$$\sum_{i,\varepsilon} (-1)^{1+\varepsilon}(-1)^{i-1}\int_{I_i^\varepsilon} \omega_{1\ldots\hat{i}\ldots p}[x^1,\ldots,(x^i=\varepsilon),\ldots,x^p]dx^1\ldots d\hat{x^i}\ldots dx^p$$

$$= \sum_{i,\varepsilon} (-1)^{i+\varepsilon}\int_{I_i^\varepsilon} i^*\omega^{p-1} = \int_{\partial I^p} i^*\omega^{p-1}$$

7.3 Stokes' Theorem for Cubical Singular Chains.

(1) Singular Cubes. Let $\emptyset$ be a C^∞ mapping into a C^∞ M: $\emptyset$: $I^p \to M^n$ ($p \leq n$) which extends to a C^∞ map of a neighborhood $U(I^p) \subset R^p$ into M. We say that $\emptyset$ is degenerate if rank $d\emptyset$ is less than p. Moreover, $\emptyset_1$ and $\emptyset_2$ are equivalent if $\emptyset_1 - \emptyset_2$ is degenerate. Then a singular cube, σ^p, is an equivalence class of nondegenerate maps $\emptyset$.

The map $\emptyset$ induces a map $\emptyset \circ i$: $I_i^\varepsilon \to M$, which is a singular cube F_i^ε of dimension p-1 called an oriented face of σ^p. By definition the boundary of σ^p, namely $\partial\sigma^p$, is:

$$\partial\sigma^p = \sum_{i,\varepsilon} (-1)^{i+\varepsilon} F_i^\varepsilon$$

(2) Singular chains. Let $\sigma_1^p, \ldots, \sigma_q^p$ be any finite set of singular cubes of dimension p. Then the formal sum

$$c^1\sigma_1^p + c^2\sigma_2^p + \ldots + c^q\sigma_q^p,$$

where the c's are integers, is called an (integral) singular cubical chain, C^p.

By definition, its boundary, ∂C^p

$$\partial C^p = \sum_i c^i \partial\sigma_i^p$$

The set of singular chains is an additive group, called the group of chains.

Example: $\partial\sigma^p$ is a p-1 dimensional singular chain.

From this definition it follows that $\partial\partial C^p = 0$. See Exercise 3.

(3) Integration over Singular Chains.

Definition. If ω^p is a form on M, then

$$\int_{\sigma^p} \omega^p = \int_{I^p} \emptyset^* \omega^p$$

Remark. It should be verified that if $\emptyset_1$ and $\emptyset_2$ are equivalent, then

$$\int_{I^p} \emptyset_1^* \omega^p = \int_{I^p} \emptyset_2^* \omega^p$$

so that $\int_{\sigma^p} \omega^p$ is uniquely defined.

Similarly we have the definition

Definition. $\int_{C^p} \omega^p = \sum_i c^i \int_{\sigma_i^p} \omega^p$

where $C^p = \sum_i c^i \sigma_i^p$

If ω^{p-1} is a form on M, then $(i^* \circ \emptyset^*)\omega^{p-1}$ is a form on I_i^ε. Then we have the definitions:

Definitions.

$$\int_{F_i^\varepsilon} \omega^{p-1} = \int_{I_i^\varepsilon} (i^* \circ \emptyset^*)\omega^{p-1}$$

$$\int_{\partial\sigma^p} \omega^{p-1} = \int_{\partial I^p} (i^* \circ \emptyset^*)\omega^{p-1}$$

$$\int_{\partial C^p} \omega^{p-1} = \sum_i c^i \int_{\partial\sigma_i^p} \omega^{p-1}$$

(4) Stokes' Theorem for Chains.

Theorem 2. If ω^{p-1} is a form on a C^∞ manifold M, then

$$\int_{\partial C^p} \omega^{p-1} = \int_{C^p} d\omega^{p-1}$$

Proof.

$$\int_{\partial\sigma^p} \omega^{p-1} = \int_{\partial I^p} (i^* \circ \emptyset^*)\omega^{p-1} = \int_{I^p} d(\emptyset^*\omega^{p-1}) = \int_{I^p} \emptyset^*(d\omega^{p-1}) = \int_{\sigma^p} d\omega^{p-1}$$

Then extend linearly to chains.

7.4 Consequences of Stokes' Theorem.

(1) Definitions.

(a) ω is closed iff $d\omega = 0$
(b) ω is derived iff $\omega = d\pi$
(c) A singular chain C is a cycle if $\partial C = 0$
(d) A singular chain C is a boundary if $C = \partial D$.

From these definitions we easily derive the following results:

Theorem 3. (a) A derived form is closed.
(b) A boundary is a cycle.
(c) If ω is closed, $\int_{\partial C} \omega = 0$
(d) If ω is derived, $\int_Z \omega = 0$ where Z is a cycle.

(2) Definitions.

(a) A function $f: C^p \rightarrow R$ is a cochain with real coefficients. Its domain is the set of all p-chains on M.
(b) The coboundary, δf, of f is defined by:

$$\delta f(C^{p+1}) = f(\partial C^{p+1})$$

If f is a p-cochain, δf is a p+1-cochain.

(c) If $\delta f = 0$, f is a <u>cocycle</u>.

(d) If $f = \delta g$, f is a <u>coboundary</u>.

<u>Theorem 4</u>. $\delta\delta f = 0$. Proof left to the reader.

(3) <u>The map k</u>. We define a map $k: \omega \to f$ whose domain is the set of p-forms on M and whose range is the set of p-cochains on M, by:

$$k\omega(C^p) = f(C^p) = \int_{C^p} \omega^p$$

<u>Theorem 5</u>. If $k(\omega) = f$, then $k(d\omega) = \delta f$

<u>Proof</u>. $\delta f(C) = f(\partial C) = \int_{\partial C} \omega = \int_C d\omega$ by Stokes' Theorem.

<u>Theorem 6.</u> (a) If ω is closed, then f is a cocycle.

(b) If ω is derived, then f is a coboundary.

<u>Proof of (b)</u>. Let $\omega = d\pi$; $k: \omega \to f$ and $k: \pi \to g$

Then $f(C^p) = \int_{C^p} \omega = \int_{\partial C^p} \pi = g(\partial C^p) = \delta g(C^p)$

So $f = \delta g$

Now consider the vector space of p-forms on M over the reals. The space of closed forms is a vector subspace Z. The space of derived forms, B, is a vector subspace of Z. Define $H = Z/B$.

Similarly the cochains with real coefficients on M form a vector space over the reals. The space of cocycles is a vector subspace $\mathcal{Z}$. The space of coboundaries is a vector subspace $\mathcal{B}$ of $\mathcal{Z}$. Define $\mathcal{H} = \mathcal{Z}/\mathcal{B}$.

Theorem 7. $k: H \to \mathcal{H}$ is a homomorphism. This is the essential content of Stokes' Theorem.

Proof. Left to the reader. The only problem is to show that k is well-defined. That is, to show that if ω_1 and ω_2 are closed forms and if $\omega_1 - \omega_2$ is derived, then $k\omega_1$ and $k\omega_2$ are in the same coset in $\mathcal{Z}/\mathcal{B}$; i.e., that the cocycle $k\omega_1 - k\omega_2$ is a coboundary.

This theorem can be greatly improved. The result is the celebrated Theorem of de Rham:

Theorem 8 (de Rham). The map $k: H \to \mathcal{H}$ is (a) onto and (b) an isomorphism.

The proof of this theorem is omitted for it requires a greater knowledge of algebraic topology than we are assuming here.

An important consequence of the fact that k is an isomorphism is the corollary:

Corollary. If $d\omega = 0$ and $\int_Z \omega = 0$ for all p-cycles, Z, with integer coefficients on M, then ω is derived on M.

The proof of this corollary assumes a result from topology, namely that if $\int_Z \omega = 0$ for all integral p-cycles then $f = k(\omega)$ is a coboundary. For the needed background in linear algebra, consult H. Whitney: "On Matrices of Integers and Combinatorial Topology", Duke Mathematical Journal, Vol. 3, pp. 35-45 (1937).

For an elementary case of this corollary see Theorem 7, Chapter 2.

7.5 Poincaré Lemma. This lemma is a local version of part of the de Rham Theorem, which is indeed needed to prove the de Rham Theorem.

We know that if $\omega = d\pi$, then $d\omega = 0$. What about the converse? This is an existence theorem in partial differential equations of a type different from those considered earlier.

Theorem 9. Poincaré Lemma. Let ω^{p+1} be a form on R^n such that $d\omega = 0$. Then locally there exists a p-form π such that $d\pi = \omega$.

Proof. (1) The Cylinder Construction. Let D be a domain in R^n and I the unit interval $[0,1]$. Then the "cylinder" is $I \times D$ whose elements are pairs (t,x) where $0 \leq t \leq 1$, and $x \in D$. Now define two maps:

$$j_1: D \to I \times D \qquad \text{where} \quad j_1(x) = (1,x)$$

$$j_0: D \to I \times D \qquad \text{where} \quad j_0(x) = (0,x)$$

If ω is a form on $I \times D$, then $j_i{}^*\omega$ is a form on D.

Let ω be a p+1 form on $I \times D$. We now define the p-form $K\omega$ on D by the formulas for monomials:

$$\text{(a)} \quad K(a(t,x)dx^{i_1}\wedge\ldots\wedge dx^{i_{p+1}}) = 0$$

$$\text{(b)} \quad K(a(t,x)dt\wedge dx^{i_1}\wedge\ldots\wedge dx^{i_p}) = \left(\int_0^1 a(t,x)dt\right)dx^{i_1}\wedge\ldots\wedge dx^{i_p}$$

This is extended linearly for general forms.

(2) Next we prove the lemma:

Lemma. $K(d\omega) + d(K\omega) = j_1{}^*\omega - j_0{}^*\omega$

Proof.

(a) If $\omega = a(t,x)dx^{i_1}\wedge\ldots\wedge dx^{i_{p+1}}$, then $K\omega = 0$, $dK\omega = 0$.

On the other hand,

$$d\omega = \frac{\partial a}{\partial t}\, dt \wedge dx^{i_1}\wedge\ldots\wedge dx^{i_{p+1}} + \text{terms free of } dt$$

$$K d\omega = \left(\int_0^1 \frac{\partial a}{\partial t}\, dt\right) dx^{i_1}\wedge\ldots\wedge dx^{i_{p+1}} + 0$$

$$= [a(1,x) - a(0,x)]dx^{i_1}\wedge\ldots\wedge dx^{i_{p+1}}$$

But $j_1{}^*\omega = a(1,x)dx^{i_1}\wedge\ldots\wedge dx^{i_{p+1}}$

$j_0{}^*\omega = a(0,x)dx^{i_1}\wedge\ldots\wedge dx^{i_{p+1}}$

So the lemma is true in this case.

(b) If $\omega = a(t,x)dt \wedge dx^{i_1}\wedge\ldots\wedge dx^{i_p}$

$$d\omega = -\frac{\partial a}{\partial x^{i_{p+1}}}\, dt \wedge dx^{i_{p+1}} \wedge dx^{i_1}\wedge\ldots\wedge dx^{i_p}$$

$$K d\omega = -\left(\int_0^1 \frac{\partial a}{\partial x^{i_{p+1}}}\, dt\right) dx^{i_{p+1}} \wedge dx^{i_1}\wedge\ldots\wedge dx^{i_p}$$

Also

$$K\omega = \left(\int_0^1 a(t,x)dt\right) dx^{i_1}\wedge\ldots\wedge dx^{i_p}$$

$$dK\omega = \left(\int_0^1 \frac{\partial a}{\partial x^{i_{p+1}}}\, dt\right) dx^{i_{p+1}} \wedge dx^{i_1}\wedge\ldots\wedge dx^{i_p}$$

Finally $j_1{}^*(\omega) = 0$ and $j_0{}^*(\omega) = 0$.

So the lemma is true in all cases.

(c) Proof of the Theorem.

Definition. D is deformable to a point iff there exists a mapping.

$$\phi: I \times D \to D$$

such that $\phi(1,x) = x$; $\phi(0,x) = P \in D$.

Hence if D is deformable to a point:

$$(\phi \circ j_1)x = x; \; (\phi \circ j_0)(x) = P$$

Now let ω be a p+1 form on D. Then $\phi^*\omega$ is a p+1 form on $I \times D$, and $j_1^*[\phi^*\omega] = \omega$; $j_0^*[\phi^*\omega] = 0$.

In the formula:

$$Kd\omega + dK\omega = j_1^*\omega - j_0^*\omega$$

write $\phi^*\omega$ for ω. Then we have:

$$Kd(\phi^*\omega) + dK(\phi^*\omega) = \omega$$

or

$$K\phi^*d\omega + dK(\phi^*\omega) = \omega$$

Since $d\omega = 0$ by hypothesis, $\pi = K\phi^*\omega$ has the required property that $d\pi = \omega$. The theorem may now be phrased more precisely.

Theorem 9': Let ω^{p+1} be defined in a domain D of R^n which can be deformed into a point P. If $d\omega = 0$, there is a p-form π in D such that $d\pi = \omega$.

Remark. This is a local theorem. For its global extension see C. B. Allendoerfer and James Eells, Jr.: "On the Cohomology of Smooth Manifolds", Commentarii Mathematici Helvetici, Vol. 32, pp. 165-179, 1958. The result referred to is the second corollary on p. 177: "Any C^∞ closed p-form on M is derivable from a C^∞ (p-1)-form on M with singularities lying on an (n-p)-cycle."

7.6 Partition of Unity.

Definition. A topological space M is called paracompact iff every open covering of M has a locally finite refinement which is also an open covering of M.

A covering is locally finite iff every $P \in M$ has at least one neighborhood which intersects at most a finite number of sets in the covering.

The following theorems are proved in texts on general topology.

Theorem 10. If M is compact, then it is paracompact.

Theorem 11. If M is Hausdorff and paracompact, then every open covering of M has a closed locally finite refinement which covers M.

Definition. A family $\mathcal{F}$ of C^∞ functions $f: M \to R^+$ (nonnegative reals) is called a partition of unity on M iff for each $P \in M$ $\sum_{f \in \mathcal{F}} f(P) = 1$. A partition of unity is called locally finite iff for each $P \in M$ there exists $U(P)$ such that all but a finite number of $f \in \mathcal{F}$ vanish on $U(P)$. A partition of unity is subordinated to an open cover of M iff for each $f \in \mathcal{F}$ there exists a set U in the cover such that $f = 0$ in the complement of U.

Theorem 12. Let M be a Hausdorff, paracompact space with a locally finite covering with coordinate neighborhoods U. Then there exists a C^∞, locally finite partition of unity subordinated to this cover.

<u>Lemma 1</u>. On R^1 let $b > a$. Then there exists a nonnegative C^∞ function f on R^1 such that in $[-a,a]$ $f(x) = 1$ and outside $(-b,b)$ $f(x) = 0$.

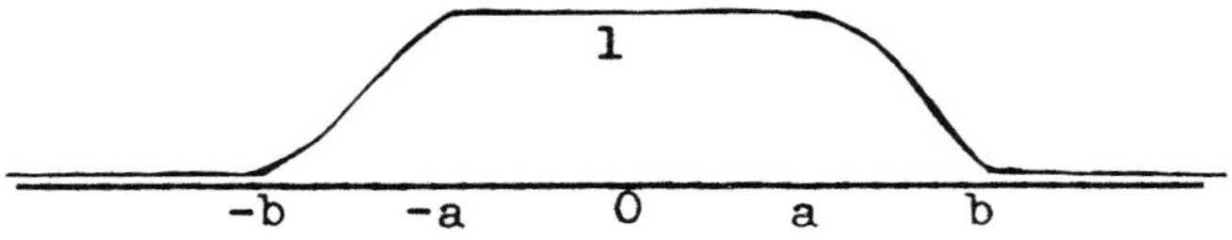

Let

$$g(x) = \begin{cases} e^{-\frac{1}{x^2}} & x > 0 \\ 0 & x \leq 0 \end{cases}$$

This is a C^∞ function as we have seen. Then define

$$h(x) = \frac{g(x)}{g(x) + g(a-x)}$$

Then $h(x)$ is C^∞, non-negative, and

$$h(x) = \begin{cases} 1 & x \geq a \\ 0 & x \leq 0 \end{cases}$$

Finally define

$$f(x) = h(x + b)h(b - x)$$

which has the desired properties.

<u>Lemma 2</u>. In R^n let $\overline{I}_a$ be a closed cube defined by $-a \leq x^i \leq a$ and let I_b be an open cube defined by $-b < x^i < b$ where $b > a$. Then there exists a nonnegative C^∞ function F on R^n which has the value 1 on $\overline{I}_a$ and the value 0 on the complement of I_b.

Proof. Let $F(x^1,\dots,x^n) = f(x^1)\times\dots\times f(x^n)$ where $f(x)$ is the function constructed in lemma 1.

Proof of the Theorem. By hypothesis we have a locally finite set of coordinate neighborhoods U_α each of which contains a closed set $\overline{V}_\alpha$ such that $\{\overline{V}_\alpha\}$ forms a locally finite cover of M.

$$\text{Let}\quad \phi_\alpha: \begin{cases} I_b \to U_\alpha \\ \overline{I}_a \to \overline{V}_\alpha \end{cases} \quad \text{be a } C^\infty \text{ homeomorphism.}$$

Let F be the function of Lemma 2, and define:

$$F_\alpha = \begin{cases} \phi_\alpha^{-1*} F \text{ on } U_\alpha. \\ 0 \text{ on the complement of } U_\alpha. \end{cases}$$

Then F_α have the properties:

(a) F_α is C^∞ on M, and is nonnegative.

(b) F_α is zero on the complement of U_α.

(c) F_α is 1 on $\overline{V}_\alpha$.

(d) F_α is different from zero in at most a finite number of neighborhoods U of the cover.

Now define

$$f_\alpha = \frac{F_\alpha}{\sum_\alpha F_\alpha}$$

The sum in the denominator is a finite sum. Also $\sum_\alpha f_\alpha = 1$.

Hence $\{f_\alpha\}$ is the required partition of unity.

7.7 Change of Variables in a Multiple Integral. In its usual form this theorem is as follows:

Theorem 13. Let A be a compact n-dimensional subset of R^n and $\phi: R^n \to R^n$ a one-to-one function defined in a neighborhood of A with a Jacobian of constant sign μ. Let $f: \phi(A) \to R$ be an integrable function. Then

$$\mu \int_{\phi(A)} f = \int_A (f \circ \phi)\, J(\phi)$$

where $J(\phi)$ is the Jacobian of ϕ; or

$$\int_{\phi(A)} f = \int_A (f \circ \phi)\,|J|$$

Since any n-form ω on R^n can be written as

$$\omega = f\, dx^1 \wedge \ldots \wedge dx^n$$

The theorem can be restated

$$\mu \int_{\phi(A)} \omega = \int_A \phi^* \omega$$

where on both sides we use the standard orientation of R^n.

The proof of this theorem is postponed to Section 7.11.

7.8 Integration on Manifolds. Our object in this section is to define the oriented integral $\int_M \omega$ where ω is an n-form and M is a connected, oriented, compact n-dimensional C^∞ manifold. Alternatively we can remove the restriction that M be compact and assume that $\omega = 0$ outside some compact subset of M.

(1) Regular Cubes and Their Orientation.

Definition. A regular cube, σ^n, is a C^∞ diffeomorphism of I^n into M whose Jacobian is not zero at any point of I^n and which can be extended to a C^∞ diffeomorphism in a neighborhood of I^n in R^n. The image $|\sigma^n|$ of σ^n is then an n-dimensional subset of M.

If $|\sigma^n|$ overlaps two of the given coordinate neighborhoods, U_α and U_β, where $U_\alpha \cap U_\beta \neq \emptyset$, two Jacobians, J_α and J_β are defined. Since $J[(\bar{\emptyset}^\alpha)^{-1} \circ \bar{\emptyset}^\beta]$ is positive, J_α and J_β have the same sign and so the sign of $J(\sigma^n)$ is well-defined.

If σ^n has local coordinates $y^1,\ldots,y^n$, its ordered set of tangent vectors $(\partial/\partial y^1,\ldots,\partial/\partial y^n)$ at each point has an orientation in terms of the fundamental basis $\left(\frac{\partial}{\partial x},\ldots,\frac{\partial}{\partial x^n}\right)$ of U_α which is given by the sign of J_α. Since this sign depends only on σ^n, we then define the orientation of σ^n by

$$\mu(\sigma^n) = \begin{cases} +1 & \text{if the Jacobian of } \sigma^n \text{ is positive} \\ -1 & \text{if the Jacobian of } \sigma^n \text{ is negative} \end{cases}$$

We can then define the orientation of M as follows:

<u>Definition</u>. The orientation of M is a cochain defined on the regular cubes by the function $\mu(\sigma^n)$.

(2) <u>Definition of the Integral of ω over M</u>. Now let M be covered by a finite set of images of regular n-cubes $|\sigma^n|_\alpha$. Let f_α be a partition of unity subordinated to this cover. Finally define

$$\int_{M,\mu} \omega = \sum_\alpha \mu_\alpha \int_{\sigma_\alpha} f_\alpha \omega$$

Then $\int_{M,\mu} \omega$ is called the integral of ω over M relative to the orientation μ. It depends on the cover $\{|\sigma^n|_\alpha\}$, and the partition of unity f_α.

<u>Theorem 14</u>. $\int_{M,\mu} \omega$ is independent of the choice of the cover and of the chosen partition of unity.

<u>Proof</u>. As usual in such proofs, we consider two covers and their associated partitions of unity:

(a) $|\sigma_\alpha|$, f_α and (b) $|\rho_\beta|$, g_β.

<u>Lemma</u>. $\mu_\alpha \int_{\sigma_\alpha} f_\alpha g_\beta \omega = \mu_\beta \int_{\sigma_\beta} f_\alpha g_\beta \omega$

<u>Proof</u>.
By definition

$$\mu_\alpha \int_{\sigma_\alpha} f_\alpha g_\beta \omega = \mu_\alpha \int_{I^n} \sigma_\alpha^*(f_\alpha g_\beta \omega)$$

and

$$\mu_\beta \int_{\rho_\beta} f_\alpha g_\beta \omega = \mu_\beta \int_{I^n} \rho_\beta^*(f_\alpha g_\beta \omega)$$

Both integrands are zero unless $|\sigma_\alpha| \cap |\rho_\beta| \neq \emptyset$, which we assume. Let $A_\alpha \subset I^n$ be the set such that $|\sigma_\alpha(A)| = |\sigma_\alpha| \cap |\rho_\beta|$

$B_\beta \subset I^n$ be the set such that $|\rho_\beta(B)| = |\sigma_\alpha| \cap |\rho_\beta|$

Then

$$\mu_\alpha \int_{\sigma_\alpha} f_\alpha g_\beta \omega = \mu_\alpha \int_{A_\alpha} \sigma_\alpha^*(f_\alpha g_\beta \omega)$$

and

$$\mu_\beta \int_{\rho_\beta} f_\alpha g_\beta \omega = \mu_\beta \int_{B_\beta} \sigma_\beta^*(f_\alpha g_\beta \omega)$$

Define $\emptyset: A_\alpha \to B_\beta$ such that where it is defined

$$\sigma_\alpha = \rho_\beta \circ \emptyset$$

Then

$$\mu_\alpha = \mu_\beta \mu_\emptyset$$

By the theorem on change of variables in a multiple integral:

$$\mu_\beta \int_{B_\beta} \rho_\beta^*(f_\alpha g_\beta \omega) = \mu_\emptyset \mu_\beta \int_{A_\alpha} (\emptyset^* \circ \rho_\beta^*)(f_\alpha g_\beta \omega)$$

$$= \mu_\alpha \int_{A_\alpha} \sigma_\alpha^*(f_\alpha g_\beta \omega)$$

This proves the lemma.

Now

$$\sum_\beta \mu_\alpha \int_{\sigma_\alpha} f_\alpha g_\beta \omega = \mu_\alpha \int_{\sigma_\alpha} f_\alpha \omega$$

So

$$\sum_\alpha \sum_\beta \mu_\alpha \int_{\sigma_\alpha} f_\alpha g_\beta \omega = \sum_\alpha \mu_\alpha \int_{\sigma_\alpha} f_\alpha \omega = \int_{M,\mu} \omega \quad \text{(using } \sigma_\alpha \text{ and } f_\alpha)$$

Similarly

$$\sum_\beta \sum_\alpha \mu_\beta \int_{\rho_\beta} f_\alpha g_\beta \omega = \sum_\beta \mu_\beta \int_{\rho_\beta} g_\beta \omega = \int_{M,\mu} \omega \quad \text{(using } \rho_\beta \text{ and } g_\beta)$$

Since the left sides are equal by the lemma, the right sides are also equal.

7.9 Manifolds with Boundary.

(1) Preliminaries. This idea is a modification of the concept of a manifold in accordance with the following definition.

<u>Definition</u>. Let N be an n-dimensional, oriented, C^∞ manifold. Then a compact manifold M with boundary ∂M is a closed, compact subset of N. The points of $M - \partial M$ are interior points and those of ∂M are boundary points.

From the C^∞ structure of N it follows that $M - \partial M$ is also a C^∞ manifold on which we can define regular cubes σ^n. We further require that ∂M be an imbedded submanifold of M. To achieve this, we introduce a collection of modified regular cubes on N.

Let K^n be the subset of R^n defined by $0 < x^\alpha < 1$ for $\alpha = 1,\ldots,n-1$ and $0 < x^n < 2$. Then I^n is the subset of K^n on which $x^n < 1$, and the face I_n^1 of I^n is the hyperplane $x^n = 1$ of K^n. Then a modified regular cube, ρ^n, is a C^∞ diffeomorphism of K^n into N whose Jacobian is not zero at any point and which can be extended to a C^∞ diffeomorphism in a neighborhood of K^n in R^n such that its restrictions to I^n and I_n^1 have the following properties:

(a) $\rho^n | I^n$ is a regular cube τ^n in $M - \partial M$.

(b) $\rho^n | I_n^1$ is a regular cube σ^{n-1} in ∂M.

Note that σ^{n-1} is contained in $\partial\tau^n$ and that it appears in the formula for $\partial\tau^n$ with the sign $(-1)^{n+1}$.

Finally, we assume that $M - \partial M$ can be covered by a finite collection of regular cubes $|\sigma_\alpha^n|$ and $|\tau_\beta^n|$ such that $|\sigma_\beta^{n-1}|$ cover ∂M.

(2) <u>Partitions of Unity</u>. Shortly we shall need a partition of unity subordinated to this cover. We do this in the usual way except that on K^n we choose the function F of Section 7.5 as in the figure:

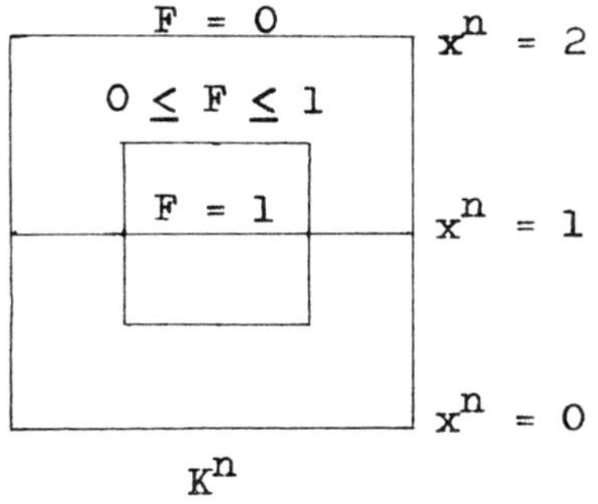

Thus we can construct a partition of unity on a neighborhood of N containing M which is thus a partition of unity on M. Let the functions of this partition be f_α on σ^n_α, f_β on τ^n_β. This induces a partition of unity on ∂M whose functions on σ^{n-1}_β are $g_\beta = f_\beta | \sigma^{n-1}_\beta$.

(3) <u>Orientation of ∂M</u>. The orientation μ of N imposes an orientation on the regular cubes σ^n and ρ^n (and hence on τ^n). If $\sigma^{n-1} = \rho^n | I^1_n$ we define its orientation ν by

$$\nu(\sigma^{n-1}) = (-1)^{n+1}\mu(\rho^n)$$

The orientation so induced on ∂M is called the orientation of ∂M coherent with that of M.

This can be interpreted in terms of vectors as follows. Let $e_1,\ldots,e_n$ be the usual basis on K^n. Then under $\rho^n(K^n) \to N$ these map into vectors $X_1,\ldots,X_n$ where $X_1,\ldots,X_{n-1}$ are tangent to ∂M and X_n is an "outer vector". The ordered set $(X_1,\ldots,X_n)$ thus has the orientation μ of ρ^n. Now the orientation of I^1_n as a portion of ∂I^n is $(-1)^{1+n}$ and so we want the orientation of $(X_1,\ldots,X_{n-1})$ to be $(-1)^{1+n}\mu$. We can accomplish this by saying that $(X_1,\ldots,X_{n-1})$ has the same orientation as the set $(X_n,X_1,\ldots,X_{n-1})$ which is $(-1)^{n+1}\mu$.

Thus
$$\upsilon(\sigma^{n-1}) = (-1)^{n+1}\mu(\rho^n).$$

7.10 Stokes' Theorem on Manifolds with Boundary.

Theorem 15. Let M be a compact manifold with boundary ∂M, μ an orientation on M, υ the coherent orientation on ∂M, and ω an $n-1$ form on M. Then $i^*\omega$ is a form on ∂M (where $i\colon \partial M \to M$ is the inclusion map) and

$$\int_{\partial M,\upsilon} i^*\omega = \int_{M,\mu} d\omega$$

Remark. We can remove the hypothesis that M be compact by assuming that $\omega = 0$ outside a compact subset of M.

Proof. Consider the finite cover of $M - \partial M$ with regular cubes $|\sigma^n_\alpha|$ and $|\tau^n_\beta|$ and the associated cover of ∂M with (σ^{n-1}_β) as described in Section 7.9. We also use the partition of unity described in Section 7.9 with functions f_α on σ_α; f_β on σ_β, and $g_\beta = f_\beta|\sigma^{n-1}_\beta$ on σ^{n-1}_β.

We observe first that since $\sum_\alpha f_\alpha + \sum_\beta f_\beta = 1$, it follows that

$$\sum_\alpha df_\alpha + \sum_\beta df_\beta = 0$$

so that

$$\sum_\alpha df_\alpha \wedge \omega + \sum_\beta df_\beta \wedge \omega = 0$$

Hence

$$\sum_\alpha \int_M df_\alpha \wedge \omega + \sum_\beta \int_M df_\beta \wedge \omega = 0$$

This implies that

$$(1) \quad 0 = \sum_\alpha \mu_\alpha \int_{\sigma_\alpha} df_\alpha \wedge \omega + \sum_\beta \mu_\beta \int_{\tau_\beta} df_\beta \wedge \omega$$

Now by definition

$$(2) \quad \int_{M,\mu} d\omega = \sum_\alpha \mu_\alpha \int_{\sigma_\alpha} f_\alpha d\omega + \sum_\beta \mu_\beta \int_{\tau_\beta} f_\beta d\omega$$

Adding (1) and (2) we obtain

$$(3) \quad \int_{M,\mu} d\omega = \sum_\alpha \mu_\alpha \int_{\sigma_\alpha} d(f_\alpha \omega) + \sum_\beta \mu_\beta \int_{\tau_\beta} d(f_\beta \omega)$$

Hence by Stokes' Theorem on cubes:

$$(4) \quad \int_{M,\mu} d\omega = \sum_\alpha \mu_\alpha \int_{\partial\sigma_\alpha} f_\alpha \omega + \sum_\beta \mu_\beta \int_{\partial\tau_\beta} f_\beta \omega$$

From the properties of the partition of unity f_α is zero on $\partial\sigma_\alpha$ and f_β is zero on $\partial\tau_\beta$ except on σ_β^{n-1} where $f_\beta | \sigma_\beta^{n-1} = g_\beta$. Recall also that σ_β^{n-1} occurs in the formula for $\partial\tau_\beta$ with the sign $(-1)^{n+1}$.

Hence from (4) we have that

$$\int_{M,\mu} d\omega = 0 + \sum_\beta \mu_\beta (-1)^{n+1} \int_{\sigma_\beta^{n-1}} g_\beta \omega$$

$$= \sum_\beta \upsilon_\beta \int_{\sigma_\beta^{n-1}} g_\beta \omega$$

$$= \int_{\partial M,\upsilon} i^* \omega$$

<u>Remark</u>. This theorem can be extended to "manifolds with boundaries and corners". That is, ∂M need not be C^∞ everywhere. All that is required is that it be C^∞ except for a

nondense subset of lower dimension.

7.11 <u>Proof of Change of Variable Theorem for Multiple Integrals</u>.
Notice that our proof that $\int_M \omega$ is independent of the cover requires the use of the change of variables theorem (Theorem 13), and that we used this in our proof of Stokes' Theorem. Now we use Stokes' Theorem to prove the change of variables theorem! To avoid circularity of reasoning, we use an induction on dimension.

(1) <u>Change of Variables Theorem in R^1</u>.

Let $\phi: R^1 \to R^1$ be written $y = \phi(x)$.
Then from elementary calculus

$$\int_a^b f(y)dy = \int_{\phi^{-1}(a)}^{\phi^{-1}(b)} (f \circ \phi)(x)\cdot\phi'(x)dx$$

or

$$\int_{\phi(c)}^{\phi(d)} f(y)dy = \int_c^d (f \circ \phi)(x)\cdot\phi'(x)dx$$

If $I = [c,d]$ then $[\phi(x),\phi(d)]$ has the standard orientation if $\phi'(x) > 0$ and the opposite orientation if $\phi'(x) < 0$. Let $\phi(I) = \mu[\phi(c),\phi(d)]$ where μ is the sign of $\phi'(x)$. Then we have

$$\mu\int_{\phi(I)} f(y)dy = \int_I (f \circ \phi)(x)\,\phi'(x)dx$$

Thus in this dimension we have

$$\mu\int_{\phi(A)} \omega = \int_A \phi^*\omega \qquad \text{where} \quad \omega = f(y)dy$$

Thus if ω is a one-form and M is a one-manifold, $\int_M \omega$ is independent of the cover.

(2) First Use of Stokes Theorem. Consider a region $D \subset M^2$ with boundary ∂D; and a 2-form ω defined on M^2 in a neighborhood of D. Since we are in the top dimension $d\omega = 0$ and so $\omega = d\pi$ in D (since D is assumed to be contractible). We are in the following position:

(a) $\int_{\partial D} \pi$ is defined independently of the cover on ∂D as a result of step (1).

(b) The integral $\int_D \omega$ is also defined, but we do not know that it is independent of the cover of D, for we do not have the change of variable theorem in two-dimensions.

(c) Stokes' Theorem $\int_{\partial D} \pi = \int_D \omega$ is proved.

(d) Hence from (c), $\int_D \omega$ is independent of the cover of D. From this we shall derive the change of variable theorem in 2-dimensions.

(3) Change of Variable Theorem in R^2. Let D be covered by the images of each of two regular cubes σ_α and σ_β with orientations μ_α and μ_β, and let ω be a two-form on M^2 in a neighborhood of D. Then from (2) $\int_D \omega$ is independent of the cover, and so

$$\int_D \omega = \mu_\alpha \int_{|\sigma_\alpha| \cap D} \omega = \mu_\beta \int_{|\sigma_\beta| \cap D} \omega$$

Recall that $\sigma_\alpha = \phi_\alpha(I^2)$ and $\sigma_\beta = \phi_\beta(I^2)$. Thus $\phi_\alpha^{-1} D$ and $\phi_\beta^{-1} D$ are subsets of I^2. By definition

$$\mu_\alpha \int_{|\sigma_\alpha| \cap D} \omega = \mu_\alpha \int_{\phi_\alpha^{-1} D} \phi_\alpha^* \omega$$

and

$$\mu_\beta \int_{|\sigma_\beta| \cap D} \omega = \mu_\beta \int_{\phi_\beta^{-1} D} \phi_\beta^* \omega$$

Let $\psi_{\alpha\beta}: I^2 \to I^2$ with Jacobian of sign $\mu_{\alpha\beta}$ be such that $(\phi_\beta \circ \psi_{\alpha\beta} \circ \phi_\alpha^{-1})$ = identity on D.

Thus $$\mu_\beta \times \mu_{\alpha\beta} \times \mu_\alpha = 1$$

Then $$\mu_\alpha \int_{\phi_\alpha^{-1} D} \phi_\alpha^* \omega = \mu_\alpha \int_{\phi_\alpha^{-1} D} \psi_{\alpha\beta}^* \circ \phi_\beta^* \omega$$

and

$$\mu_\beta \int_{\phi_\beta^{-1} D} \phi_\beta^* \omega = \mu_\alpha \mu_{\alpha\beta} \int_{(\psi_{\alpha\beta} \circ \phi_\alpha^{-1}) D} \phi_\beta^* \omega$$

From the above derivation the two right hand sides are equal, and so

$$\mu_{\alpha\beta} \int_{\psi_{\alpha\beta}(\phi_\alpha^{-1} D)} \phi_\beta^* \omega = \int_{\phi_\alpha^{-1} D} \psi_{\alpha\beta}^* (\phi_\beta^* \omega)$$

With a change of notation: $A = \phi_\alpha^{-1} D$, $\bar{\omega} = \phi_\beta^* \omega$; $\psi_{\alpha\beta} = \bar{\phi}$, $\mu_{\alpha\beta} = \mu$, this becomes

$$\mu \int_{\bar{\phi}(A)} \bar{\omega} = \int_A \bar{\phi}^* \bar{\omega}$$

which is the change of variable theorem in R^2.

(4) Continuation. Now we have Stokes' Theorem on M^3, where the boundary integral is independent of the cover. Hence the three-dimensional integral is independent of the cover, and we get the Change of Variable Theorem in R^3. The process continues by induction.

7.12 Integration of Forms on Submanifolds. Let M^p be a differentiable manifold which is a submanifold of a differentiable manifold N^n where $p < n$. Then there is a C^∞ map $\emptyset\colon M \to N$ with Jacobian of rank p.

Let M have a boundary ∂M and $i\colon \partial M \to M$ be the inclusion map. Then if ω^{p-1} is a $(p-1)$-form on N, $\emptyset^*\omega^{p-1}$ is a form on M and $(i^* \circ \emptyset^*)\omega^{p-1}$ is a form on ∂M. Thus from Stokes' Theorem we have the formula:

$$\int_{\partial M}(i^* \circ \emptyset^*)\omega^{p-1} = \int_M d(\emptyset^*\omega^{p-1}) = \int_M \emptyset^* d\omega^{p-1}$$

7.13 Riemannian Metric.

Definition. A Riemannian metric at a point $P \varepsilon M$ is a symmetric, bilinear mapping from pairs of vectors at P to the reals, denoted by $\langle X_P, Y_P \rangle_P \,\varepsilon\, R$ such that $\langle X_P, X_P \rangle_P \geq 0$ and $\langle X_P, X_P \rangle_P = 0$ iff $X_P = 0$. A global Riemannian metric on M is a set of real-valued C^∞ functions of P, denoted by $\langle X, Y \rangle$ such that if X and Y are C^∞ vector fields on M, $\langle X, Y \rangle_P = \langle X_P, Y_P \rangle_P$ is a Riemannian metric at P. In local coordinates $\langle X, Y \rangle = g_{ij}X^iY^j$ where (g_{ij}) is a positive definite matrix.

Theorem 16. A Hausdorff, paracompact C^∞ manifold M admits a globally defined Riemannian metric.

<u>Proof</u>. Consider the locally finite cover U_α with coordinate neighborhoods and in U_α let $<X,Y>_\alpha = \delta_{ij} X^i Y^j$. Then $f_\alpha <X,Y>_\alpha$ is globally defined and $f_\alpha <X,Y>$ is positive in $\overline{V}_\alpha$ for $X \neq 0$.

Finally define

$$<X,Y>_P = (\sum_\alpha f_\alpha <X,Y>_\alpha)_P$$

This has the required properties.

<u>Remarks</u>. (1) The corresponding conjecture concerning the global existence of a metric tensor of signature $+ - - -$ on M^4 is false.

(2) By means of the above construction we have formed a globally defined tensor of type $(0,2)$ whose local coordinates are g_{ij}. These components are symmetric in i and j and the matrix (g_{ij}) is positive definite. As in Chapter 4 we denote its inverse by (g^{ij}) whose elements are the components of a globally defined tensor of type $(2,0)$. We denote $\det(g_{ij}) \neq 0$ by $\underline{g}$.

(3) The existence of a Riemannian metric permits us to establish on M a metric in the usual sense. Define the distance $PQ = \underset{\{C\}}{\text{glb}} \int_P^Q \left|\frac{dx}{dt}\right| dt$ where C is a curve joining P and Q, and $\left|\frac{dx}{dt}\right|$ locally has the expression:

$$\left|\frac{dx}{dt}\right| = \left(g_{ij}(x)\frac{dx^i}{dt}\frac{dx^j}{dt}\right)^{1/2}$$

This satisfies the requirements of a metric and makes M a metric space.

7.14 Integration on Riemannian Manifolds. A Riemannian manifold M^n is an n-dimensional differentiable manifold with the additional structure of a Riemannian metric as described in Section 7.13. Such manifolds are abstractions from the hypersurfaces of R^n discussed in Chapter 4. In this section we shall assume that our manifolds are oriented.

(1) Volume form. If g_{ij} are the local components of the metric tensor in M, we have seen in Chapter 4 that $g = \det(g_{ij})$ determines the local volume element dV $dV = \sqrt{g}\,dx^1 \ldots dx^n$, and that if x and $\bar{x}$ are overlapping coordinate systems

$$\sqrt{g} = \sqrt{\bar{g}}\,|J| \quad \text{where} \quad J = \det(\partial\bar{x}/\partial x).$$

These facts motivate us to define the general volume form, dV, on M as follows.

First we consider the quantities $\varepsilon_{i_1 \ldots i_n}$ of Chapter 5 which are defined by:

$$\varepsilon_{i_1 \ldots i_n} = \begin{cases} +1 \text{ if } i_1, \ldots, i_n \text{ is an even permutation of } 1 \ldots n \\ -1 \text{ if } i_1, \ldots, i_n \text{ is an odd permutation of } 1 \ldots n \\ 0 \text{ otherwise} \end{cases}$$

If x and $\bar{x}$ are overlapping coordinate systems, we have seen (Chapter 5, exercise 14) that

$$\varepsilon_{i_1 \ldots i_n} \frac{\partial \bar{x}^{i_1}}{\partial x^{j_1}} \cdots \frac{\partial \bar{x}^{i_n}}{\partial x^{j_n}} = J\,\varepsilon_{j_1 \ldots j_n}$$

where J is the Jacobian determinant $J = \det(\partial\bar{x}/\partial x)$. Because of the factor J in this expression, $\varepsilon_{i_1 \ldots i_n}$ are not components of a tensor of type $(0,n)$ as might be supposed from their expressions. But since J is assumed to be positive and $\sqrt{g} = \sqrt{\bar{g}}\, J$, it follows that

$$\sqrt{\bar{g}}\, \varepsilon_{i_1 \ldots i_n} \frac{\partial \bar{x}^{i_1}}{\partial x^{j_1}} \cdots \frac{\partial \bar{x}^{i_n}}{\partial x^{j_n}} = \sqrt{g}\, \varepsilon_{j_1 \ldots j_n}$$

Hence $\sqrt{g}\, \varepsilon_{j_1 \ldots j_n}$ are components of a tensor of type $(0,n)$.

Thus we may define the n-form

$$dV = \frac{1}{n!} \sqrt{g}\, \varepsilon_{i_1 \ldots i_n} dx^{i_1} \wedge \ldots \wedge dx^{i_n} = \sqrt{g}\, dx^1 \wedge \ldots \wedge dx^n$$

as a global, nowhere zero form on M. We assume that the set $(dx^1, \ldots, dx^n)$ is positively oriented relative to the orientation of M. This form dV is called the volume form. If M is compact, $\int_M dV$ is the volume of M.

The integral of a function f on M can now be defined to be $\int_M f dV$.

(2) Hypersurfaces. The metric tensor on M induces a metric tensor on a hypersurface H^{n-1} of M as follows. A hypersurface H^{n-1} is defined by a set of local maps $\phi\colon R^{n-1} \to M$ whose Jacobian matrix is of rank n-1. Let ϕ be expressed in local coordinates by $x^i = f^i(u^\alpha)$ $\alpha = 1, \ldots, n-1$. Then

$$ds^2 = g_{ij} dx^i dx^j = g_{ij} \frac{\partial x^i}{\partial u^\alpha} \frac{\partial x^j}{\partial u^\beta} du^\alpha du^\beta = h_{\alpha\beta} du^\alpha du^\beta$$

Thus $h_{\alpha\beta} = g_{ij} \frac{\partial x^i}{\partial u^\alpha} \frac{\partial x^j}{\partial u^\beta}$ defines a Riemannian structure on H^{n-1}.

To discuss the tangent and normal vectors to H^{n-1} in M^n we proceed as in Section 4.13. But here we must take into account the fact that the inner product of two vectors V^i and W^i tangent to M is given by

$$< V,W > = g_{ij} V^i W^j$$

Also, to each contravariant vector V^i there is associated a covariant vector $V_j = g_{ij}V^i$ such that $g^{ij}V_j = V^i$. Thus the length of V^i, namely $|V|$ can be written alternatively as:

$$|V|^2 = g_{ij}V^iV^j = V_iV^i = g^{ij}V_iV_j.$$

For each α, $T^i_\alpha = \frac{\partial x^i}{\partial u^\alpha}$ are vectors of M tangent to H^{n-1}, and the set $\{T^i_\alpha\}$ are independent. We now define the covariant vector V_i to be:

$$V_i = \varepsilon_{ii_1\ldots i_{n-1}} T_1^{i_1} \ldots T_{n-1}^{i_{n-1}}$$

Then V_i is normal to H^{n-1} since for each α:

$$V_i T^i_\alpha = \varepsilon_{ii_1\ldots i_{n-1}} T^i_\alpha T_1^{i_1} \ldots T_{n-1}^{i_{n-1}} = 0$$

We can find the length of V as in Section 4.13. Let $\Delta = \varepsilon_{i_1i_2\ldots i_n} V^{i_1} T_1^{i_2} \ldots T_{n-1}^{i_{n-1}}$. This is a determinant which we write symbolically as $\begin{vmatrix} V \\ T_1 \\ \cdot \\ \cdot \\ \cdot \\ T_{n-1} \end{vmatrix}$. From our definition of V, it follows that $\Delta = |V|^2$; and so, among other things, $\Delta > 0$. If $g = \det(g_{ij})$, we find that $\Delta \times g$ is the determinant

$$\begin{vmatrix} V^i g_{i1} & \cdots & V^i g_{in} \\ T_1^i g_{i1} & \cdots & T_1^i g_{in} \\ \cdot & & \cdot \\ \cdot & & \cdot \\ \cdot & & \cdot \\ T_{n-1}^i g_{i1} & \cdots & T_{n-1}^i g_{in} \end{vmatrix}$$

and

$$\Delta \times g \times \Delta^T = \begin{vmatrix} |V|^2 & 0 \\ 0 & h_{\alpha\beta} \end{vmatrix}$$

So $|V|^4 g = |V|^2 h$ where $h = \det(h_{\alpha\beta})$

Thus

$$|V| = \frac{\sqrt{h}}{\sqrt{g}}$$

and the unit vector N_i in the direction of V_i has the expression:

$$\sqrt{h}\, N_i = \sqrt{g}\, \varepsilon_{ii_1 \ldots i_{n-1}} T_1^{i_1} \cdots T_{n-1}^{i_{n-1}}$$

(3) <u>Forms of Degree n-1 Defined by Vector Fields</u>. Let $P^i(x)$ be a contravariant vector field on M. Then we define its

associated (n-1)-form ω by:

$$\omega = \frac{1}{(n-1)!} P^i \varepsilon_{i i_1 \ldots i_{n-1}} \sqrt{g}\, dx^{i_1} \wedge \ldots \wedge dx^{i_{n-1}}$$

Moreover

$$d\omega = \frac{\partial}{\partial x^i}(P^i \sqrt{g})\, dx^1 \wedge \ldots \wedge dx^n$$

$$= \frac{1}{\sqrt{g}} \frac{\partial}{\partial x^i}(P^i \sqrt{g})\, dV$$

If we recall from Exercise 19 of Chapter 5 that the divergence, div P, equals $\frac{1}{\sqrt{g}} \frac{\partial}{\partial x^i}(P^i \sqrt{g})$, we can finally write write:

$$d\omega = (\text{div } P) dV$$

(4) Gauss' Divergence Theorem. This is a very useful result in applied mathematics which is essentially a reformulation of Stokes' Theorem. Let M be a manifold with boundary so oriented that $(dx^1, \ldots, dx^n)$ has the orientation of M and $(du^1, \ldots, du^{n-1})$ the orientation of ∂M. Our assumption that ∂M is coherently oriented relative to M is equivalent to the statement that the determinant Δ of (2) above:

$$\Delta = |N, T_1, \ldots, T_{n-1}|$$

is positive. This implies that N is the outward normal to ∂M.

Let P^i be a contravariant vector field on M and consider the (n-1)-form:

$$\omega = \frac{1}{(n-1)!} P^i \varepsilon_{i i_1 \ldots i_{n-1}} \sqrt{g}\, dx^{i_1} \wedge \ldots \wedge dx^{i_{n-1}}$$

This induces a form $i^*\omega$ on ∂M given by:

$$i^*\omega = P^i \varepsilon_{ii_1 \cdots i_{n-1}} \sqrt{g}\, T_1^{i_1} \cdots T_{n-1}^{i_{n-1}}\, du^1 \wedge \cdots \wedge du^{n-1}$$

$$= (P^i N_i)\sqrt{h}\, du^1 \wedge \cdots \wedge du^{n-1}$$

$$= (P^i N_i)\, dV(\partial M)$$

Since from (3) above, $d\omega = (\operatorname{div} P)\,dV(M)$ it follows from Stokes' Theorem that we have proved the theorem of Gauss:

<u>Theorem 17</u>. If M is a Riemannian manifold with boundary ∂M, P^i is a contravariant vector field on M, and N is the unit outer normal to ∂M, then

$$\int_{\partial M} (P^i N_i)\, dV(\partial M) = \int_M (\operatorname{div} P)\, dV(M)$$

where the integrals are taken with coherent orientations.

7.15 <u>Classical Forms of Stokes' Theorem</u>.

(a) <u>Green's Theorem in the plane</u>. Let $\omega = P\,dx + Q\,dy$, D any domain in R^2 whose boundary is a differentiable curve (or set of curves) C. Then

$$\int_C P\,dx + Q\,dy = \iint_D \frac{\partial Q}{\partial x} - \frac{\partial P}{\partial y}\, dx\,dy$$

<u>Corollary</u>. A line integral $\int P\,dx + Q\,dy$ along a curve joining points A and B is "independent of the path" iff $\frac{\partial Q}{\partial x} - \frac{\partial P}{\partial y} = 0$.

(b) <u>Stokes' Theorem for Surfaces in R^3</u>. Let $\omega = P\,dx + Q\,dy + R\,dz$ be defined in R^3 and

S a submanifold of R^3 bounded by a curve C. Then by 7.12

$$\int_C (i^* \circ \phi^*)\omega = \iint_S \phi^* d\omega$$

or

$$\int_C P\frac{dx}{dt} + Q\frac{dy}{dt} + R\frac{dz}{dt}\, dt =$$

$$\iint_S \left[\left(\frac{\partial R}{\partial y} - \frac{\partial Q}{\partial z}\right)\left(\frac{\partial(y,z)}{\partial(u,v)}\right) + \left(\frac{\partial P}{\partial y} - \frac{\partial R}{\partial x}\right)\left(\frac{\partial(z,x)}{\partial(u,v)}\right) + \left(\frac{\partial Q}{\partial x} - \frac{\partial P}{\partial y}\right)\left(\frac{\partial(x,y)}{\partial(u,v)}\right)\right] du \wedge dv$$

$$= \iint_S (\text{curl}) \underset{\sim}{P}) \cdot \underset{\sim}{N}\, d\sigma$$

where $d\sigma$ is the element of area on S, and N is the normal such that C and S are coherently oriented.

(c) Gauss's Divergence Theorem in R^3. Let $\omega = P dy \wedge dz + Q dz \wedge dx + R dx \wedge dy$ be defined in R^3. S a closed surface in R^3 with interior D. Then

$$\iint_S \omega = \iiint_D \left(\frac{\partial P}{\partial x} + \frac{\partial Q}{\partial y} + \frac{\partial R}{\partial z}\right) dx \wedge dy \wedge dz$$

or

$$\iint_S \underset{\sim}{P} \cdot \underset{\sim}{N}\, d\sigma = \iiint_D \text{div}\, \underset{\sim}{P}\, dV$$

where N is the outer normal and S and D are coherently oriented.

References

Cartan, H.: Differential Calculus, Houghton Mifflin, Boston, 1971.

Cartan, H.: Differential Forms, Houghton Mifflin, Boston, 1970.

Eisenhart, L. P.: Riemannian Geometry, Princeton, 1949.

Flanders, H.: Differential Forms, Academic Press, New York, 1963.

Fleming, W. H.: Functions of Several Variables, Addison-Wesley, Reading, 1965.

Flett, T. M.: Mathematical Analysis, McGraw-Hill, New York, 1966.

Lang, S.: Introduction to Differentiable Manifolds, Interscience, New York, 1962.

Milnor, J.: Morse Theory, Princeton, 1963. See Part II: A Rapid Course in Riemannian Geometry.

Spivak, M.: Calculus on Manifolds, Benjamin, New York, 1965.

Sternberg, S.: Lectures on Differential Geometry, Prentice-Hall, Englewood Cliffs, 1964.

Taylor, A. E.: Advanced Calculus, Ginn, New York, 1955.

Veblen, O.: Invariants of Quadratic Differential Forms, Cambridge, 1933.

Warner, F. W.: Foundations of Differentiable Manifolds and Lie Groups, Scott, Foresman, Glenview, Illinois, 1971.

References (Continued)

Willmore, T. J.: Differential Geometry, Oxford, 1959.

Woll, J. W.: Functions of Several Variables, Harcourt, Brace and World, New York 1966.

Exercises

1. Write out the details and thus verify the formula for $d\omega^{p-1}$ of Section 7.2, namely that

$$d\omega^{p-1} = \left[\sum_i (-1)^{i-1} \frac{\partial \omega_{1\ldots\hat{i}\ldots p}}{\partial x^i}\right] dx^1 \wedge \ldots \wedge dx^p \mu$$

2. Show that if the map ϕ of Section 7.3 is degenerate, then

$$\int_{I^p} \phi^* \omega^p = 0$$

3. Prove that for singular chains, Section 7.3, $\partial\partial\sigma^p = 0$.

4. Prove that for cochains, Section 7.4, $\delta\delta f = 0$.

5. Show that for 0-forms (i.e. functions) defined on 1-chains, Stokes' Theorem is the same as the Fundamental Theorem of Calculus.

6. Prove that a derived form is closed.

7. Prove that a boundary is a cycle.

8. Prove that if ω is closed, $\int_{\partial C} \omega = 0$

9. Prove that if ω is derived, $\int_Z \omega = 0$ where Z is a cycle.

10. If α and β are closed forms, prove that $\alpha \wedge \beta$ is closed. If in addition β is derived, then $\alpha \wedge \beta$ is derived.

11. Prove that the map $k: H \rightarrow \mathcal{H}$ of Section 7.4 is a homomorphism.

12. Prove that every closed 1-form on the sphere S^2 is derived.

13. Let $\omega = y dx + (x + z) dy + y dz$
   (a) Show that $d\omega = 0$
   (b) Find π such that $d\pi = \omega$
   (c) Is π unique? If not, find an expression for all such forms π.

14. Integrate the 2-form

$$\omega = dx \wedge dy + 2dy \wedge dz + z^2\, dz \wedge dx$$

over the top half $(z \geq 0)$ of the ellipsoid

$$\frac{x^2}{4} + \frac{y^2}{9} + \frac{z^2}{16} = 1$$

(Do not use brute force, but be clever.)

15. Let $\phi: R^3 \to R^3$ be given by

$$x = r \sin\theta \cos\phi$$

$$y = r \sin\theta \sin\phi$$

$$z = r \cos\theta$$

Find $\phi^*(dx \wedge dy \wedge dz)$ and thus find the element of volume of R^3 in spherical polar coordinates.

16. Let z be a complex variable $x + iy$,
$f(z) = u(x,y) + i\, v(x,y)$, and

$$\int_C f(z)dz = \int_C u dx - v dy + i \int_C v\, dx + u\, dy$$

where C is a simple closed curve in R^2.

16. (Continued) Prove that $\int_C f(z)dz = 0$ for all curves C if

$$\frac{\partial u}{\partial y} = -\frac{\partial v}{\partial x} \text{ and } \frac{\partial u}{\partial x} = \frac{\partial v}{\partial y}.$$

These are the Cauchy-Riemann equations of Complex Variable Theory.

17. In Sections 7.13 and 7.14 we proved the existence of a nowhere zero n-form on any orientable manifold. Prove the converse: If there exists a nowhere zero n-form on a differentiable manifold M^n, then M^n is orientable.

18. Let $\varepsilon^{i_1 \dots i_n}$ have the same values as $\varepsilon_{i_1 \dots i_n}$. Then prove that $\frac{1}{\sqrt{g}} \varepsilon^{i_1 \dots i_n}$ are the components of a tensor of type $(n,0)$.

19. Let $P_{i_1 \dots i_{n-1}}$ be the coefficients of an $(n-1)$-form on a differentiable manifold. Impose a Riemannian metric and define:

$$P^i = \frac{1}{(n-1)!} \frac{1}{\sqrt{g}} \varepsilon^{ii_1 \dots i_{n-1}} P_{i_1 \dots i_{n-1}}$$

Thus derive Stokes' Theorem from Gauss' Theorem. Do not restrict your result to Riemannian manifolds.

20. In Gauss' Theorem put $P^i = \frac{\partial F}{\partial x^j} g^{ij}$ where F is a a function. Write $\frac{\partial F}{\partial x^j} N_i g^{ij} = \frac{\partial F}{\partial n}$

20. (Continued) Then prove that:

$$\int_{\partial M} \frac{\partial F}{\partial n}\, dV(\partial M) = \int_M \nabla^2 F\, dV(M)$$

(b) By putting $P^i = u \frac{\partial v}{\partial x^j} g^{ij}$ where u and v are functions, prove "Green's first identity":

$$\int_{\partial M} u \frac{\partial v}{\partial n}\, dV(\partial M) = \int_M \frac{\partial u}{\partial x^i} \frac{\partial v}{\partial x^j} g^{ij} + u \nabla^2 v\, dV(M)$$

(c) Hence prove "Green's second identity":

$$\int_{\partial M} \left(u \frac{\partial v}{\partial n} - v \frac{\partial u}{\partial n}\right) dV(\partial M) = \int_M (u \nabla^2 v - v \nabla^2 u)\, dV(M)$$

INDEX

Affine
    approximation to a function, 8
    function, 8
Algebra
    exterior, 131
    multilinear, 123
Allendoerfer, C. B., 190
Analytic function, 6
Arc length, 91, 98
Atlas, 142

Banach space, 30
Birkhoff, G., 51
Boundary, 185
    of I^p, 180
    orientation of, 199

C^p, C^∞ and C^ω functions, 6
Cartan, H., 214
Cauchy-Riemann equations, 218
Chain rule, 9
Chains
    cubical singular, 183
        integration over, 184
Change of variables in a multiple integral
    theorem, 193
        proof, 202
Chart, 97, 142
Christoffel symbols, 103
Closed form, 185
Coboundary, 185
Cochain, 185
Cocycle, 186
Coddington, E. A., 51
Component functions, 2
Connection, 102, 168
    coefficients of, 102, 168
Continuity, 2
Contraction, 130
Coordinate system, 142
Cotangent vector, 152
Covariant derivative, 105, 168
    directional, 107, 168
    exterior, 171
    on a manifold, 168

Covectors, 131
    basis for, 133
    exterior product of, 134
Critical point, 70, 165
Cubes
    Euclidean, 180
        boundary of, 180
    regular, 194
        orientation of, 195
    singular, 183
Curvature of a
    curve in R^n, 93
        first, second, etc., 94
    curve on a hypersurface
        geodesic, 110
        normal, 110
    hypersurface
        Gaussian, 112, 119
        mean, 112
        principal, 109, 112
        total, 112
Curve, 89
    arc length of, 91, 98
    differentiable, 91
    Frenet equations of, 92
Cycle, 185

de Rham, theorem of, 47, 187
Derivative
    covariant, 105
        directional, 107
    directional in R^n, 9
    exterior
        abstract definition, 157
        invariant definition, 158
        local definition, 156
    partial, 4
Derived form, 185
Determinant, 139 ex. 13-18
Differentiability of functions, 3
    continuous, 5
Differentiable manifold, 142
Differential, 4, 164
Differential equations, 23
    approximate solutions of, 32
    differentiability of solutions of, 32, 34
    existence theorem, 23
        contraction map proof, 29
        Picard proof, 24
        unicity of solution, 28
    linear, 34

Differential equations
partial, 39, 47
Frobenius theorem, 47
total, 82, 159
Differential forms
closed, 185
definition, 154, 155
derived, 185
exterior derivative of, 156
integration of, 180, 194
Directional derivative, 9
Divergence, 140 ex. 19
Gauss' Theorem, 211, 213
Dual vector space, 123
basis for, 123
double, 127
Dyadic product, 127

Eells, J., Jr., 190
Einstein, A.
metric, 100
summation convention, 98
Eisenhart, L. P., 214
Euclidean cube, 180
boundary of, 180
Euclidean space, 1
basis, 1
inner product, 1
norm, 1
standard topology, 1
Exterior
algebra, 131
derivative, 156
product, 134
Extreme values of a function, 70

Flanders, H., 214
Fleming, W. H., 214
Flett, T. M., 214
Form
differential, see Differential forms
first fundamental, 100
second fundamental, 101
Frenet equations of a
curve, 92
solution of, 95
hypersurface, 100, 102
Frobenius theorem
first form (differential equation), 47
second form (tangent vector fields), 79

Frobenius theorem (Continued)
    third form (total differential equations),
        general, 159, 178 ex. 22
        preliminary, 82
Functions
    analytic, 6
    component, 2
    continuity, 2
    differentiability of, 3, 143
    differential of, 4
        norm of, 4

Gauss, K. F.
    and Codazzi equations, 113
    curvature, 112
    divergence theorem, 211, 213
    spherical image, 118
    Theorema Egregium, 114
Geodesic, 110
    curvature, 111
Gordon, W. B., 88
Gradient of a function, 39
Green's
    first identity, 219
    second identity, 219
    theorem in the plane, 212

Homomorphism $k: H \to \mathcal{H}$, 187
Homotopy, 41
Hypersurface
    in R^n, 97
    in M^n, 208

Implicit function theorem, 58
Ince, E. L., 51
Integrability conditions, 39
Integration of
    forms
        on manifolds, 180, 194, 195
        on Riemannian manifolds, 207
        on submanifolds, 205
    functions
        on R^n, 18
        on Riemannian manifolds, 208
Interior product, 137 ex. 5
Inverse function theorem, 54

Jacobian, 5

Lagrange multipliers, 72
    second derivative test, 72
Lang, S., 214
Laplacian, 140 ex. 20
Levinson, N., 51
Lie bracket, 79, 152
Lie derivative, 177 ex. 17
Line integration, 40
    independence of the path, 40, 45
Linear differential equations, 34
Lipschitz condition, 23

Manifold, 68
    differentiable, 142
    imbedded, 68
    immersed, 68
    with boundary, 197
        coherent orientation of boundary, 199
Mappings, 163
    critical point of, 165
    differential of, 164
    nondegenerate, 167
Maximum value of a function, 70
    constrained, 71
    second derivative test, 70, 72
Mean value theorem
    $R^n \to R^1$, 12
    $R^n \to R^m$, 14, 21 ex. 14, 15, 16
Miller, K. S., 51
Milnor, J., 214
Minimum value of a function, 70
    constrained, 71
    second derivative test, 72
Multilinear algebra, 123
Murray, F. J., 51

Newns, W. F., 147
Nomizu, K., 97
Normal vectors to a
    curve, 93
        binormal, 94
        principal, 94
    hypersurface, 69, 209
    submanifold of R^n, 68

Orientable manifold, 174
Orientation of
    boundary, 199
    manifold, 173

Orientation of (Continued)
    vector space, 135

Paracompact space, 191
Parallel
    displacement, 171
    vectors in Euclidean space, 149
Partial differential equations, 39, 47
Partial derivatives, 4
Partition of unity, 191
Picard method, 24
Poincaré lemma, 188
Positive definite matrix, 70
Product of tensors, 128
    exterior, 134
    interior, 137 ex. 5
Projection functions, 2
    map, 150

R^n, 1
Riemannian
    manifold, 207
        hypersurface of, 208
        volume form on, 207
    metric, existence of, 205
Rota, G. C., 51

Sard's theorem, 165
Simply connected region, 41
Spherical image, 118
Spivak, M., 214
Sternberg, S., 214
Stokes' theorem for
    cubical, singular chains, 185
        consequences of, 185
    I^p, 181
    manifolds with boundary, 200
    surfaces in R^3, 212
Submanifolds of R^n, 64
Symmetric matrix, diagonalization of, 75

Tangent bundle, 149
    Euclidean space, 149
    orientation of, 174
    projection map, 150
Tangent space, 144
    basis for, 147
Tangent vectors to a
    curve, 90
        unit, 92

Tangent vectors to a (Continued)
    differentiable manifold, 144, 147
    hypersurface, 100, 209
    submanifold of R^n, 98
Taylor, A. E., 214
Taylor's theorem, 15
    with Lagrange remainder, 16, 17
Tensor, 99
    components, 126
    contraction, 130
    contravariant, 127
    covariant, 127
    definition, 124
    field, 155
    first fundamental, 99
    law of transformation, 104, 126
    metric, 100
    product of vector spaces, 128, 130, 138 ex. 8-12
    second fundamental, 101
Thomas, T. Y., 51
Torsion of a curve, 94
Total differential equations, 82, 159
Trajectory, 90

Veblen, O., 214
Vector
    cotangent, 152
    dual, 123
    field, 151
    normal, see Normal vectors
    tangent, see Tangent vectors
Volume
    form, 207
    of a hypersurface, 115, 118
    of a Riemannian manifold, 208

Walker, A. G., 147
Warner, F. W., 214
Whitney, H., 187
Willmore, T. J., 214
Woll, J. W., 214